CHUBB GUIDE TO HOME SECURITY

PROTECT YOUR HOME AND PROPERTY FROM FIRE AND THEFT

CHUBB GUIDE TO HOME SECURITY

Michael M. West

Peter Carter

KEY PORTER BOOKS

Canadian Cataloguing in Publication Data

West, Michael M.
Chubb guide to home security

ISBN 1-55013-152-4

1. Dwellings — Security measures. 2. Burglary protection. I. Carter, Peter.
II. Title.

TH9745.D85W48 1989 643'.16 C89-094082-7

Typesetting: Southam Business Information & Communications Group Inc.
Printed and bound in Canada
Illustration pages 25, 28, 51, 59, 92 Dean Kalmantis

Key Porter Books Limited
70 The Esplanade
Toronto, Ontario
Canada M5E 1R2

89 90 91 92 93 5 4 3 2 1

CONTENTS

CHUBB GUIDE TO HOME SECURITY

INTRODUCTION

THE SECURITY BLANKET

This book is written for two types of people. If you have never had to face a burglary or a fire in your home or apartment, it will tell you not only what you can do to prevent such disasters but also how to prepare for them and thereby minimize their upsetting effects. If you've already been a victim of robbery or fire, the book contains information that will lessen the chances of it ever happening again.

As anyone who has been robbed or had their home broken into will tell you, there's a certain naiveté in buying a home and not reinforcing it, particularly when you contrast the cost of a home today with the low cost of security. Not only will a break-enter-and-theft or a fire deprive you of goods that you have worked hard to obtain but it can make you more cynical and less trusting of other people. If we are to eradicate the crime of break-and-enter in particular, we must have a complete trust in our neighbors and associates—a return to good old-fashioned community values. It's difficult to maintain an attitude of honesty and trust if your own experiences make you cynical.

Unfortunately, the statistics show an upswing in break-and-enter crimes in Canada, and it is now the most common criminal offence. On average, a Canadian home is broken into every two and a half minutes. Between 1978 and 1987, there was a 31.5 percent increase in break-and-enters across the country. And almost 80 percent of these remain unsolved. The rate varies from place to place, but generally, there is just under one break-in per year for every 100 people.

BREAK AND ENTER OFFENCES

Year	Residential	Business	Other	Total
1979	164,192	93,292	38,953	296,437
1984	209,923	101,856	45,133	356,912
1985	213,443	98,859	44,442	356,744
1986	217,675	101,536	44,929	365,140
1987	217,503	99,172	44,863	361,538
1988	214,173	100,697	44,382	359,252

Source: Statistics Canada, Canadian Centre for Justice Statistics.

- 30% increase in residential break-ins between 1979 and 1988 (10 years), compared to 8% increase in business burglaries for same period.
- For every business broken into, two houses are burglarized.
- Homes in metropolitan and city areas are more prone to burglary attack. For example, break and enter of houses increased 13.5% in 1988 to 9,998 (8,806 in 1987) in Metropolitan Toronto.

If you haven't been struck by a burglar, statistics indicate that you know somebody who has, whether you live in Halifax, Orillia or Yellowknife. Across this country, there is a good chance that one in every 38 homes will be broken into this year. And according to some authorities, that estimate might be low. In 1987 alone, thieves robbed your fellow Canadians of at least

$80 million in valuables. The most popular targets for burglars include items found in virtually every home – stereo equipment, jewelry, furs, coin collections, cameras, VCRs, watches and cash.

This book considers the two greatest threats to home security: break-enter-and-theft, and fire. Chapters One through Five tell the story of an actual break-and-enter in a major Canadian city and, with the shaken homeowner, consider the various options available for making your home as secure as possible from break-ins. Also, Chapter Five takes you beyond break-and-enters to areas of concern such as car security, potential fraud scams, cottage security—essentially how to incorporate security-mindedness into your everyday life.

Chapters Six and Seven are devoted to fireproofing your home as extensively as possible. Chapter Eight examines the role of insurance in home security and tells you what to do if disaster should strike. It's much too easy to underestimate your insurance needs, and Chapter Eight describes how you can make sure you don't find out you have too little insurance too late. Finally, the Afterword provides a handy security checklist, a quick way for you to start determining your personal security needs.

As your home is probably the largest single investment you will make in your lifetime, it only makes sense to protect it as well as you possibly can. The time spent reading this book and thinking about your home's vulnerability will pay off tremendously, whether or not you are ever directly affected by crime or fire. Forewarned is forearmed, and you'll be pleased to find that a little common sense and a few fairly inexpensive precautions around your home can offer you peace of mind and greater home security.

CHAPTER ONE

NO PLACE LIKE HOME

"I had quite an enjoyable weekend up at the cottage – fished, took it easy. It was so nice I stayed an extra day and came back very late, around 11:30 or midnight. The timer was on, so the lights in the living room were on when I got home. As I pulled up, I parked right in front of the house. I looked up and saw the light in my bedroom on, and I thought immediately, 'Charlene' – that's my ex-wife – 'must be here, she must have come over for something.' She's the only one who has a key.

"Then as I walked toward the front door and noticed the door open, I thought, 'Oh, well she's just arrived or she's stopped in to get something.' The screen door was closed but the front inside door was open. So I opened the screen door and I remember hesitating, taking a deep breath, and noticing that the light in the den, where I keep the VCR and TV and stuff, was on. And that was very peculiar. And I noticed the blinds on the patio windows over the back yard were askew. 'Something's wrong here,' I thought. I was standing at the entrance to the house, and my eyes darted to the rack of stereo equipment, and

I saw all the stuff there — or so I thought. Then I walked in and went straight to the back toward the blind.

"I yelled. I yelled 'Charlene,' then 'Who's here?' and there was no answer. And the house was cold. At that time of year, I hadn't turned the furnace on, and I could feel a breeze. And then I saw the glass on the floor and the jagged broken window. It all sunk in. It all sort of came home.

"It's just this sickening feeling. Blood just drains from your face, you get this sort of empty sensation. There's a sense of being victimized, a feeling of being almost punched.

"I got to the den, where the window was broken, and I have a cabinet that the VCR and amplifier sat on, with a pull-out shelf. The shelf was pulled out and nothing was on it. And by this time, I was starting to get real angry. He took the VCR, the new amplifier, the predictable stuff. Of course the portable CD player was gone.

"Then I raced upstairs. I went into the bedroom and everything was turned over and the bedside-table drawers had been pulled out. You could tell he really worked fast.

"There's a panicky feeling, then you get a little more rational. You're really angry, sort of excited, and you want to tell someone. So I phoned the police, and that was about 20 minutes after I got home. Then while you wait, you just get this surge of adrenaline. You go around saying to yourself, 'I'll get the low-life that did this.' "

There are few experiences in life as painful as having your home, your private sanctuary, violated by thieves. And that's what happened to Alan. His experiences, his feelings and his changed life are typical of innocent people who get robbed.

Alan's neighborhood is like countless other Canadian communities, with a mix of families and young professional couples living side by side with homeowners of their parents' age. People park on the streets in front of their homes, and neighbors eventually get to know one another. Like other Canadians, the residents along Alan's street seldom barricade their doors with arsenals of locks and anti-theft devices. But as Alan found out, after he was struck twice, things in every part of this country are changing. Not only do break-and-enter victims improve their home hardware, they usually undergo a change of attitude as well.

The man who sold the house to Alan had himself been the victim of a break-in and as a result had barricaded the outside basement door. "The only thing I did about security was when I went away on trips," Alan says. "I bought a couple of timers for the lights.

"Although I wasn't even considering the possibility of getting broken into when I was buying the house, the previous owner went to great pains to point out that at considerable expense he had installed a pick-proof lock on the front door. It uses this eccentric strange-looking key that looks like no other key on earth, and I guess maybe it is pick-proof. And in fact, as locks go it is a very good lock. There's a way you can lock it so if somebody's inside the house and doesn't have the key, they can't get out."

The trip that forced Alan to re-examine his house and his attitude toward home security took place in the fall of 1988. "It was Labour Day weekend, apparently a big weekend for robberies. So the cops told me after the fact. Not that I would have done anything, even if I had known. I've gone away lots of Labour Day weekends and the only thing I had was a timer on the light in the living room. . . ."

From a Burglar's Perspective

Call it second-guessing, Monday-morning quarterbacking or just plain procrastination. It always seems easier to figure out how to solve a problem after the damage has been done. By now, there have been enough burglaries in this country that you can learn from them.

Generally, break-and-enter thieves are young men or juveniles on the lookout for quick cash. The majority of those who are caught maintain they are either on drugs or alcohol or looking for money to buy drugs or alcohol. Most aren't dangerous — unless they're cornered. They are not highly skilled thieves, and yet the cleverest of the bunch are never detected. Quite often, the doors they break through are already open or have such inadequate locks that all they have to do is push hard. They are after easily salable items like VCRs, stereos, jewelry, watches, furs or, better yet, straight cash.

When a burglar reports for work, the first thing he does is check if anyone is home. If not, he then finds a way to get into your house. Once inside, he establishes an easy escape route, cases the living-room booty and heads for the master bedroom, where the search begins. Then, about 10 minutes later, he can leave through the escape route and be on his way to sell your goods. Typically, he will get about 10 percent of their original value. The only thing a thief knows for sure is that he does not want to be caught.

If anyone knows how to deter burglars, it should be the burglars themselves. Recently, the American Southland Corporation, owner of the 7-Eleven convenience-store chain, wanted to redesign their stores to make them more robbery-proof. Wisely, the company selected as the architect of the plan a man named Ray D. Johnson, who, before he worked for 7-Eleven, served 25 years in California prisons for robbery and burglary. On Johnson's recommendations, the company redesigned 60 outlets in Southern California. After his ideas were implemented, the 7-Eleven people watched the theft rate in the renovated stores drop by 30 percent.

But you probably don't have the same security needs as the local convenience store. And you might not have access to an experienced thief who can help you determine what kind of challenge your particular property presents to burglars. Therefore you ought to start assessing your home security needs by looking at your property through the eyes of a criminal. Then, not only do you realize how valuable your property and belongings are but you also learn how easy your home is to break into and rob.

When you move into a new dwelling, the last thing on your mind is how your home appears to potential thieves. And when you spend your hard-earned money to purchase appliances, home-entertainment equipment, watches, jewelry or collectibles for your hobby, you are not expected to think about how attractive the item might be to thieves. In fact, that attraction does not have to be a consideration at all, providing your home is secure. But the point is, if you are like most law-abiding, trusting citizens, you have gathered a surprisingly large number of goods over the years that thieves would love to get their hands on, and your dwelling is extremely vulnerable. Indeed, like most law-abiding citizens, you don't think it will happen to you. You don't have bars on your windows.

Assessing your vulnerability as well as your worth is an educational, entertaining and rewarding exercise. So let's start by getting a thief's-eye view of your home. With notebook in hand, take two slow walks around your property—one at night, and one during the morning after the neighbors have gone to work. These are the times when most burglaries take place. As you notice shortcomings in your home's security, jot them down. Then, if you think you should rectify them, refer to the appropriate chapter in this book. You may want to refer to the Home Security Quiz in the Afterword as you evaluate your home's security.

First of all, take a good, honest look at the kind of street you live on and ask yourself some questions:

- How well do you know your neighbors?
- Would your next-door neighbors know enough about you to consider calling the authorities if they noticed a stranger doing something unusual in your back yard? Also, if you get to know them a little, you can ask your neighbors to give your home that lived-in look when you are on vacation.
- Have any of your neighbors been the victims of thieves? That might give you some idea of how likely you are to be hit.
- In most of the families on your street, are both husband and wife working away from home? If so, your home is particularly attractive to the nine-to-five burglar.
- Ask the police about the crime statistics of your neighborhood, because these figures can indicate your own chances of becoming a burglary statistic.
- Find out how long it would take the police to get to your home in an emergency. If the nearest detachment is several

miles away, as it sometimes is in rural areas, there is extra cause for concern.

That done, take a close look at your home itself.

• How close is your dwelling to the road?

• Can neighbors see your street address if they want to report something? Large numbers, 6 inches (15 cm) high and preferably well lit, placed on both the front and the back of the house, are deterrents.

• Would a burglar have to cross a lot of property from the street or back alley to get to your house?

• If you were a burglar, how much time would it take you to case the house from the street, cross the property, gain entrance, do your dirty work, exit, get off the property and away from an area? And could all that be done without attracting the attention of suspicious neighbors? There's only one thing a thief hates more than lost time, and that's time lost in trying to explain to curious neighbors what he's doing on your property.

• From the outside, can a thief tell if you have taken security precautions? Window bars, visible from the exterior, are one indication that the occupant is security-conscious. The same thing applies to a guard dog. If a home bears a Neighbourhood Watch or Operation Identification sticker, thieves will be less likely to risk getting caught. Better yet, an alarm system connected to a central security station promises even more deterrence, providing you advertise the fact, with a sticker on the door or windows, to thieves who might be window-shopping in your area. The two most effective proven roadblocks for thieves are clearly centrally monitored alarm systems and guard dogs. Police say that if you can deter a would-be thief for just three minutes, you will likely prevent a break-in.

• Do your windows, doors, fences, gates and hinges look secure and well maintained? A thief is less likely to try robbing a home where the occupants appear to be conscientious about their property. In a burglar's mind, householders who are careless about repairs are probably also careless about security precautions — they might even be the kind of people who leave money lying around inside. Loose-looking doors that are clearly in a state of disrepair, torn screens and broken windows are open invitations to a thief.

• How well lit is your house? Brightly illuminated homes, with lights both outside and in, are less likely targets than dark, seemingly vacant dwellings. Not only do lights send out a signal that somebody's home but in the case of poorly illuminated property, a thief could stand in an unlit doorway and work on breaking the lock on your door without being seen. In the war against break-and-enter artists, deliberate, well-placed lighting is one of your strongest allies.

• When you're away for the evening, are you like millions of homeowners who leave one or two lights on inside your house? If the pattern is constant, your ploy will quickly become apparent to the prospective thief who checks out your dwelling two or three nights in a row. Think about what your home looks like when you are away from it during an evening or morning. Do you leave your curtains or blinds open or closed? Closed curtains suggest vacancy. On the other hand, open curtains provide a prowler with the chance to case your rooms from outside. Would a thief strolling down your street or back alley be able to peek through your window and know immediately that the house contains valuables? In many Canadian neighborhoods, not only in *trendier* areas, it's not unusual to stroll down a street and look right past broadly opened curtains onto a crystal chandelier.

• How many shaded areas are there around your house? Are any windows hidden by bushes or trees? Does a tree stand between a streetlight's glare and your house? Are there corridors of darkness along hedges, through gardens or beside swimming-pool fences? Wherever you see shadows, a burglar sees safe entry and exit routes.

• Do any trees provide access to your upper-story windows? Do windows sit conveniently over porch roofs or fire escapes? Is there by any chance a ladder in your yard (or stored in an unlocked shed)? Just think how inviting a ladder or a window-mounted air conditioner (which can be easily removed) looks to a thief. A burglar assessing his chances of a successful robbery would not only spot the ladder or air conditioner but he would also note things like the second-story window that you leave slightly open all the time to let the cat in and out.

• How secure are your outbuildings? Not only do garages and sheds contain such valuables as bicycles, motors and power tools but they are also home to handy items like crowbars, screwdrivers and saws, items a thief might borrow to enter your home that much more easily. What kind of lighting system does your garage have? Even if it contains absolutely nothing of value, a garage can be a perfect hiding place for a burglar waiting for the right opportunity to hit your home. If your garage is attached to the house, do you keep the connecting door locked? You must consider the security system of your attached garage as seriously as you think about one for the rest of the house.

If your outbuildings are separate from the house, how secure are they? Take note of the lighting, windows, doors, locks — the same aspects you think about when you're assessing your dwelling. Is there any way of telling from within the house that someone is in the garage or shed? Would you be able to spot an intruder from a kitchen or bathroom window? Or might you hear him if he opened the garage door? Imagine the sheer terror of discovering that your outbuildings had been entered and robbed while you were right there at home.

• Are there valuables lying around your yard? If so, you're just inviting trouble. Bicycles, lawn mowers, lawn furniture, barbecues and sporting goods top this list of easy pickings. As well, as we mentioned, when would-be burglars see such a casual approach to belongings, they naturally assume the rest of your property deserves the "sitting-duck" rating.

Finally, think about what your house looks like through a criminal's eyes while you are away on vacation:

• Do newspapers and mail pile up?

• Do your family vehicles remain in the same spot hour after hour, day after day?

• Does snow pile up on the path and the driveway — or does grass grow too high — and advertise the house's vacancy?

Now that you have looked at your home just the way a

burglar would, move in for a much closer look, searching out the weak points that, while they could never be seen by an outsider, would play a significant role in a burglar's activity. Try to imagine how a canny burglar would approach them. That way, you can conduct your personal-security survey with the worst interest of the professional thief at heart.

Of primary importance are your major points of entry — windows and doors. Do they lock? The question might seem odd until you begin a closer examination. Particularly if they live in older homes, many people live in dwellings where windows and garages simply do not have locks. Or you might be among those people who leave windows and garage doors unlocked, thereby weakening considerably your home's burglary-immune system. According to one survey, about 20 percent of burglars gain access through a window, and an unlocked window makes the thief's job that much easier. Even if all your doors and windows do have locks, the next question to answer honestly is, do you lock them, even when you're just putting out the garbage or raking leaves? Most thieves enter through doors, and about half of those doors either aren't locked or have inadequate locks, the sort that even the most amateur thief could bypass or jimmy open.

Some door locks, as you will see in Chapter Two, are easier to force open than others. Some are near windows, so a thief can smash the glass, reach in and unlatch the lock. Some people make a thief's job even simpler by hanging keys on a nail right behind the window by the door. For economical reasons, a homeowner might opt for a hollow wooden door — which a thief can punch or hammer his way through to reach the inside latch. The same applies to doors with built-in windows or sidelights, which, once broken, give handy access to the knob inside. That thought alone should be sufficient to make you carefully reconsider your locking system.

How easily can your locks be broken or bypassed? If, for instance, your door uses a popular cylinder lock with a spring-latch mechanism, is there enough space for an intruder to insert a credit card to push the latch open?

Doors, like thieves, come in many shapes and sizes. When

you examine your doors, see if the hinges are exposed to the exterior. If they are, has the manufacturer constructed the hinge carefully enough so a screwdriver-equipped thief can't remove the hinge bolt? Also, think about doors of the sliding variety, doors onto the balcony or patio. Instead of sliding, try lifting one of these doors open. You just might get an unpleasant surprise.

Chapter Two goes into further detail about door construction, window materials, lock varieties and precautions you can take for unusual entranceways like patio doors or skylights. But while you are sizing up your property's vulnerability, familiarize yourself with every possible point of entry, even those through which you would never consider crawling. How secure, for instance, are your basement windows? Often in older dwellings, basement windows at ground level and encased in the concrete foundation go untended while the homeowner confidently installs reinforced casement windows upstairs. All ground-level windows deserve extra consideration. So do skylights, windows that house air conditioners, windows that adjoin glassed-in porches with the home's living space, and pet entranceways.

For What It's Worth

As you measure your home's vulnerability, you might find yourself asking the question, "Do I have anything worth stealing?" Of course you do. Just remember that break-and-enter crimes are the work of criminals who by their own admission overestimate the potential booty from any given crime. That means that even if you don't think your valuables are worth protecting, a thief will probably disagree and enter your home hopefully. And just because a household contains little of apparent value doesn't mean the intruder is not going to wreak havoc once he's inside. Essentially, criminals don't care if you have anything worth taking. If they are determined to hit your house, they will.

Furthermore, when you tell yourself that your worldly goods are not necessarily very valuable, you are probably mistaken. After you have cased the physical situation and construction of your home, sit down and figure out what you own and how much of that a thief would like to make his own. Once you start, it won't take long to figure out that to a thief your net worth rivals that of Fort Knox. (See the Checklist of Household Valuables in the Afterword.)

If you have determined "through the eyes of a thief" that your home is penetrable and contains something worth stealing, it's time to look at increasing your level of home security. Later, we will discuss assessing the monetary value of your goods and how that relates to proper insurance. But for now, keep at the heart of your deliberations the notion that security and peace of mind transcend the dollar sign by light years.

CHAPTER TWO

AT THE THRESHOLD

"The guy had broken in through this small octagonal window near the back of the house. The window is sheltered by big bushes and it isn't off the ground so it's really easy. It's one of the two most vulnerable spots in the house. The other is the basement door, which I didn't give a second thought to because it had these big wooden braces on it. I have a patio door too, but apparently thieves seldom break the glass in patio doors, although they can be lifted off the track. So I installed a piece of wood that prevents that.

"There was no light at all on the north side where the window was. Plus there was this big bush, so if you could get in there behind the bush you could get in through that window and nobody would see you."

Depending on your approach, safeguarding your home against thieves can either drive you to delirium as you sit and fret about the countless methods that nefarious burglars have for outwitting your home-security measures, or you can take the easy route. With the first route, the end is sheer confusion as you worry about your regional crime statistics, your travel and work habits, what can happen if you hear a burglar in the house, or any other combination of unpleasant hypothetical situations. On the other hand, the easy route leads to a secure home and peace of mind. And the key to simple and effective home security is to remember that a thief intent on breaking into your home really has only two entranceways: windows and doors. Seldom are homes entered through holes cut in ceilings or floors. If you count features like skylights and garage entrances as windows and doors, and if you take the precaution of keeping your windows and doors in good solid condition with dependable locking mechanisms, you can take a more relaxed but effective approach to home security.

About 80 percent of housebreaking thieves enter through the door. Some gain access through doors that are actually left unlocked and even open. Others make it in past inadequate locks and leave little or no signs of forced entry. Another widely used tactic is to smash the window in or beside the door, reach in and unlatch the door from the inside. Add to this the fact that the most popular tool employed by break-and-enter culprits is the simple screwdriver, and you can quickly assess the vulnerability of your doors.

Try this test. Imagine yourself locked out of your house, with nothing but a standard screwdriver to help you break in. At first glance, you would probably say the prospects of gaining easy access through the front door — with the least amount of damage — are grim. But be completely ruthless and don't worry about how badly you would mangle the door frame, screen, hinges, lock mechanisms or nearby windows. By splitting the frame and levering the screwdriver into the latch, or by smashing a door window, breaking into your house becomes considerably easier.

This chapter will discuss in more detail the kinds of things you have to think about when securing the vulnerable doors, windows and locks in your house. Remember that most thieves aren't worried about damaging your property. All they want to do is get in, steal, and get out — fast. And they will do as much damage as necessary to accomplish their mission.

A final point. Any time you return home, whether from work or a vacation, if you sense that something is wrong or looks out of order about your house, don't go in. Go immediately to a neighbor's place and call the police from there. You never know who is in your house or what kind of dangerous reaction you might provoke.

Doors

Strong doorways will always be your best defense. When thinking about the doors on your house, keep in mind not only the front and back doors but also patio doors, inner garage doors and basement doors.

Although locked screen doors provide some initial discouragement to thieves, they are in fact the most vulnerable of entranceways. A properly operating screen door, though easily broken, will at least slow down the determined burglar. If a screen or storm door swings loosely on its hinges, or has a torn screen, a burglar will assume that the rest of the property needs maintenance as well and will probably be an easy mark. So make sure your screen doors lock the way they are supposed to. If a thief comes across a locked screen door, he is far more liable to think somebody is home. Obviously, anything that discourages or slows down a thief improves your chances of avoiding loss.

Main doors are another matter. The reason that so many thieves can get in through the main doors is simply that these entranceways are often less secure than they might be.

First of all, a lot of modern homes are built with standard-issue hollow wooden doors. True, they are much less expensive than a wooden door with panels, but the amount of security provided by the hollow models is minimal. In many cases, the only thing between a thief and your belongings is two sheets of thin plywood sandwiching a honeycomb framework. You can bet that when you see one of those TV police shows where the detective kicks a door in to catch the bad guys in the act, the cop is putting his foot through a hollow, not a solid, door. So if your entranceways have hollow doors, you might think about replacing them with solid doors.

Solid doors should be a minimum of 1¾ inches (4.5 cm) thick. Nevertheless, if there is a window in your door, it won't matter how strong the door is because a thief can break the window, reach inside to the doorknob and let himself in. Again, don't forget that thieves don't mind breaking things. A windowpane, even if it is in a solid wood or metal door, should never be more than 20 square inches (130 cm^2) or within 40 inches (100 cm) of the doorknob. Mailboxes are preferable to slots in the door, but if you have a mail slot in your door, it should be built so that a person reaching through cannot reach the doorknob inside. Incidentally, *never* "hide" a key in the mail slot or mailbox. That's the first place any self-respecting thief looks.

Because of the close connection between windows in doors and accessibility, you ought to give the glass in your doors more consideration than you would the glass in your main windows. Side windows, which border the main door of many homes, are important too. Although most people do a lot of looking through their door windows, it makes sense to use wire mesh in them to make them harder to break. Or you might consider installing thick glass blocks in the side windows. They're much more break-resistant than regular glass, and they range in density from the virtually opaque to the crystal clear.

If you would rather not have mesh in your door windows, you should investigate having them cut from material that is stronger than standard glass, such as Plexiglas, tempered glass, laminated glass or Lexan, a polycarbonate product. Otherwise you might want to investigate space-age window-coating materials such as Profilon, a reinforced plastic film with an adhesive backing that you apply over the inside of regular glass panes to prevent them from shattering. Profilon is less expensive than Lexan and special glass products. Plexiglas can withstand an impact three times the force that would shatter ordinary glass, and Lexan has been known to survive a blow of 100 times that. So it wouldn't matter if a thief came equipped with an ax instead of the usual screwdriver; he still wouldn't be able to smash a Lexan pane. Likewise, specially treated films like Profilon will provide a formidable barrier to a thief's break-in attempts, even if he arrives with a full supply of bricks or a sledgehammer.

None of these materials is completely burglarproof, however. Most can be cut with a drill, a saber saw or removed completely. The best they can do is decrease your chances of loss.

There is one piece of glass that you must install in your door, and that is the one in an optical viewer, which lets you peer out and get a wide-angle look at whoever is knocking. You should install optical viewers in all exterior doors that do not have windows.

While we're discussing state-of-the-art building materials, you might like to think about installing metal doors and frames, which are stronger still than solid-core wooden models. A screwdriver-equipped thief cannot hack his way around the locking mechanism of a steel door. Likewise, a steel frame is almost impossible to pry wide open. But they do have their limitations. The locking mechanism on steel doors is in many cases a simple key-in-knob system, and that's easy prey for even a semi-skilled lock picker. And the do-it-yourself homeowner will have a more difficult time installing extra locking devices in tough steel doors.

Of course, your doors could be constructed of solid metal and still remain vulnerable to the canny (or even not-so-canny) criminal. Any door is only as good as the frame that surrounds it and the lock that holds it closed. Whether you live in your own home, a condominium or an apartment, your doors must be maintained just like any other mechanical mechanism. So be on guard for deterioration. Over the years, as a building settles and changes with the environment, door frames warp. Doors don't fit their frames as snugly as they did when they were new. Sometimes, the wood around a door becomes chipped or loosened through use over the years.

A very popular trick among thieves is to slip a credit card (presumably a stolen one) into the crack between the door and the frame to pry the spring latch of a cylinder lock open. The technique is known as "loiding," in honor of the celluloid from which credit cards were once made. When we discuss locks, you will see that only the most ineffective of modern locks can be easily loided and that deadbolts are impervious to this old trick. But no door should allow a thief to insert anything, even something as thin as a credit card, between the door and the frame. More destructive but gaining in popularity among thieves is the method of simply sticking a crowbar or screwdriver between the door and the frame and hacking away until the wooden "mortise" area and frame are destroyed.

You might want to try loiding for yourself, to see how easy it can be. There's yet another effective method of bypassing a vulnerable door frame, but you probably won't feel like testing it. A determined thief who has done any planning will come prepared, even if that means carrying a car jack to help do the job. Imagine how easy it would be to lever your door frame if you installed the car jack sideways between the two sides of the door frame. Once the frame is widened a little, the entire lock mechanism is released and the door pops open. The key to preventing this method is simply strong door frames. Door frames should be solid and of one-piece construction. Not only should frames be straight, but the studs on either side of the frame around the locking mechanism, particularly in the central area, should be reinforced with wooden blocks. As stated earlier, time and the environment take their toll on any door, and the frames around your doors may be in sore need of reinforcement. If a frame is loosely attached to the stud or the concrete behind it, a thief with a crowbar can enter your home with some ease. Frames should be secured to their supporting structure with nothing less than 2- or 3-inch (5-8 cm) screws. Shorter screws can be ripped out of their holes with very little leverage.

Doors and door frames that have deteriorated through the years should be replaced. Even if you opt not to change the entire structure, you should replace or adjust the locking mechanism, which won't work as effectively on the ill-fitting door and frame.

Just as a well-prepared thief can cut through the wood around the door frame, he might also use his tools to cut into the actual door. To avoid such drastic measures on his part, you could insert long slender steel rods above and below the lock and knob mechanisms. These reinforcing bars prevent any thief cutting his way past your lock.

The manner in which a door is mounted can have a bearing on its vulnerability. Some homes have main doors that open outward, so that the hinges are exposed. It doesn't take much experience to figure out how to use a screwdriver as a lever to extract a hinge pin from its slot so you can pry the door open. If your doors do open outward, leaving the hinges exposed, you can drill small holes through the hinge and insert short pegs or screws, making removal of the pins nearly impossible. Another way of keeping hinges in place is to flatten both ends of the hinge pin so it cannot be pried out. A third alternative is to remove one of the screws that fasten the hinge to the door, insert a small peg into the hole, leaving half of it sticking out, and drill a corresponding hole into the frame for the peg to fit into when the door is closed. The peg will act as a bolt between the door and the frame, and a thief won't be able to remove the closed door even if he takes out the hinge pin.

For doors that don't have hinges, there are other means of reinforcing security. More and more homes are being constructed with attractive broad glass patio doors that slide open. Like fireplaces, under the right circumstances patio doors can be charming and useful additions to a home. But like fireplaces, if patio doors suffer from neglect or misuse, they invite trouble. Often the locks that are built into sliding patio doors are flimsy and easily broken. Also, the glass in the door is usually breakable, and thus the door is vulnerable to forced entry. So in your efforts to burglarproof your dwelling, patio doors deserve extra attention.

Consider what kind of glass is in your patio doors. Depending on how worried you are about being broken into, you might consider having the doors constructed of specially strengthened materials such as Plexiglas or Lexan, or coated with Profilon. These are expensive but practical materials.

Nonetheless, it wouldn't matter how strong the glass was if a thief were able to grab the two sides of the door and lift it out of its slots. There are several simple precautions you can take to avoid that possibility.

First of all, by drilling a couple of self-tapping screws into the upper metal track but leaving their heads slightly exposed, you can eliminate almost all the headroom that would otherwise permit someone to lift the door from its slot. A thin strip of wood or metal inserted along the track works the same way.

Merely because your patio door cannot be lifted off its track doesn't mean a thief can't use his trusty screwdriver to break the lock and then slide the door open. To prevent this, you can drill holes right through the door frame into the casement and then insert steel pins, nails or small screws into the holes. These screws or pins act as deadbolt locks and have to be withdrawn in order for you or anyone else to slide the door open. You may want to attach these pegs to a small chain or string so they don't get lost when you take them out. You may choose to purchase spring locks, which work the same way and are manufactured specifically for the job, or track grips, which are attached to the lower track and secure the sliding panels in the closed position.

Patio doors must also be secured with some sort of locking bar. One option is a Charlie Bar, installed either into the frame or onto the edge of the sliding door about 4 or 5 feet (1-1.5 m) from the bottom. It folds down horizontally and prevents the door from sliding open. Some have key-locking safety latches so they can't be raised even if a burglar slices through the glass and reaches through to remove the bar.

Another, smaller lock employs a curved length of sturdy metal that affixes to the stationary panel of the patio door and prevents a thief from either sliding the door open or lifting the door off. If your patio doors do not already have this type of lock, you can install one yourself quite easily.

Some newer homes come with large hinged patio doors that swing open instead of sliding. With these, it is important that the locking mechanisms at the center as well as at the top are sturdy. In the end, if a thief is determined to enter your home through a locked patio door, and if the door contains breakable glass, no lock will keep him out. Though patio door locks can slow him down, if he's willing to live with the broken glass, you will have to as well. Still, properly installed locks, especially when they're visible from outside, add to the secure appearance of your home, and that's sometimes enough to keep a thief away.

Finally, don't forget the garage doors. They should be either automatic or padlocked to a hole in the door-mounting track.

Door Locks

The door lock, despite being in existence since at least as far back as 1000 B.C., has yet to be perfected. But it sure has come a long way. Many homes in Canada, particularly apartments in buildings that were at one time single-family dwellings, continue to employ the simple-to-defeat key-in-knob cylinder lock with a spring-latch locking mechanism. As mentioned above, a time-tested method of opening such a lock is the familiar "loiding" with a plastic card. With many spring-latch cylinder locks in use today, loiding is a snap. And as locks have evolved, so have break-and-enter techniques. Thieves sometimes come equipped with lock-breaking devices that rival surgical tools in their complexity. Security-conscious homeowners owe themselves the peace of mind that comes from knowing that the best locks available are separating their goods from the hands of thieves.

Your household's security is only as tight as its weakest lock. This cannot be emphasized enough. As obvious as the point seems, you might have the strongest locks available on your main door but weaklings on your basement or garage entranceways.

Common sense is essential when assessing your locks. First, do you know where all the keys to all the locks are? Do each of your kids carry keys? If one of them loses one, change the locks (that is, have the cylinder rekeyed). When you move into a new home, change the locks. If you are among the unlucky Canadians who have been broken into, and especially if the thief has come in through the door, change the locks immediately. And never hide keys in obvious places such as under doormats or in mailboxes. Neglecting any one of these measures can make your home very attractive to prospective burglars.

In replacing the locks, remember that high-quality, high-security locking systems offer "key control." This means that locksmiths and general key-cutting services will refuse to duplicate keys as they would not do with older or less-secure systems. The high-security keys are marked "Do Not Copy," and householders who wish to have additional keys cut for themselves must show proof that they are the authorized owners of the key. For ultimate key control, buy a lock with a restricted keyway. With this variety, no other locksmith has keyblanks for your lock, dramatically reducing the possibility of unauthorized key duplication.

Your first task is to find out what kind of lock suits a particular door the best. A simple chain lock, or "door interviewer," provides virtually no protection, but it will let you open the front door a crack so you can talk to visitors without letting them in. If a basement door doesn't get too much use, perhaps it would be best served by an extremely secure bar or jimmy-proof lock. Either works well, though neither looks very appealing on your front door.

The most secure entranceways have some combination of lock so locks can employ several types of mechanisms. And different manufacturers offer different features for each kind of lock. Your best move is to seek out a qualified locksmith. Overall, locks should be constructed of hacksaw-proof hardened steel, and there should be some sort of protection on the outside of the door that prohibits a thief from prying or twisting off the keyway. Select a lock that uses high-security mounting hardware, including a strikeplate that must be recessed into the door frame as opposed to the surface mounted type. The larger the strikeplate, the greater the pressure it can withstand. Any lock is only as secure as the material to which it is attached. If the mounting screws are short or the wood is rotted, a thief will have an easy time breaking the door and entering your house.

Some models of lock are combinations of two or more types. The term "cylinder" means that the lock uses a key whose notches and grooves activate pins allowing the entire cylinder mechanism to be turned. Cylinders and keys can be used to activate beveled, spring-equipped latches that spring shut automatically. Spring latches can serve as the main door latch, or they can be built into a rim lock mounted on the edge of the door and jamb. In either case, spring latches, especially if used with a key-in-knob type lock offer extremely low-level security. Spring latches are most useful on interior doors, such as the bathroom. You may find a spring latch that has a small bolt installed alongside the latch. An adept thief can pry this open too. If you must install a key-in-knob latch-bolt lock,

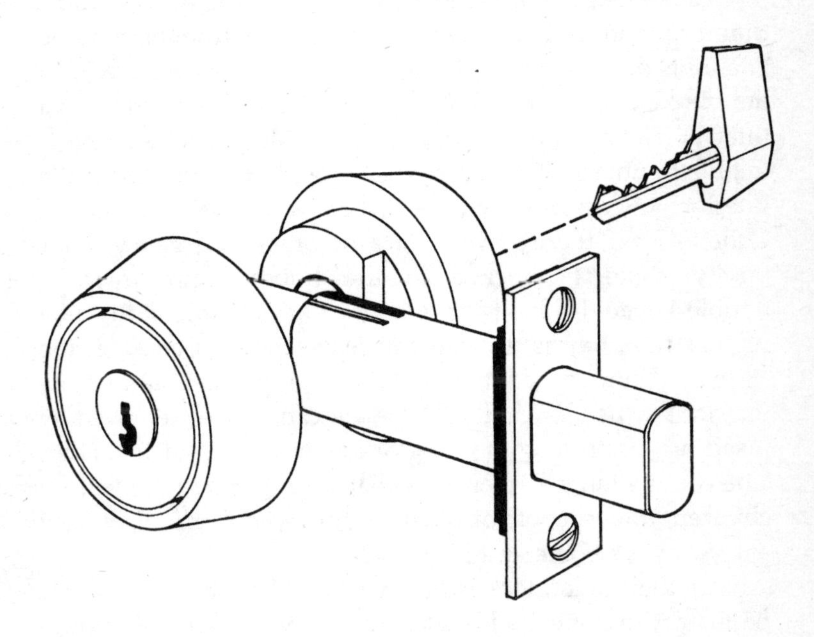

AUXILIARY DEADBOLT LOCKS

A double-cylinder deadbolt lock, with a removable thumbturn key on the inside, is the best protection for doors with windows or glass sidelights. The thumbturn key can be removed when the house is unoccupied.

choose only a high-quality lock of hardened steel. Make sure it uses the auxiliary protection provided by the bolt beside the latch to prevent loiding. However, it's better to complement the cylinder with another type of auxiliary deadbolt locking mechanism.

Deadbolt locks are more secure. They have a longer "throw," that is, they extend much farther into the door frame. Deadbolts can be mounted on the rim of the door and jamb, or installed inside the door so that the bolt emerges from the door and enters a metal strikeplate within the door frame. The first is easier to install, but you may find the second more attractive. A "jimmy-proof" rim-mounted lock employs a vertical deadbolt which unites the frame and door sections of the lock, thus providing greater security than a single horizontal bolt.

Some locks are operated by a key from the outside and a thumbturn on the inside. Some require a key from both inside and outside the house. Your security is increased with a double-keyed (keyed on both sides) deadbolt, because even if a thief breaks a window and reaches inside to the latch he is unable to unlock it. It is essential, though, that you leave the interior key in the cylinder whenever anyone is home. Otherwise, if there's ever a fire or other emergency, your family could be locked in the house. Some modern double-keyed deadbolts have interior keys that look just like regular thumbturns, so you can leave them inserted in the keyway. With any double-keyed lock, it is essential that the interior key be the "captive" type — it can only be removed by inserting and turning the exterior key in the outside cylinder. The captive inner key ensures that it cannot be removed by children, tradespeople or other visitors who might want easy entry to your home at a later date.

Another, older, lock is the mortise. Mortise locks are not popular with home builders because they cost more to install. Available with deadbolts or a combination of deadbolt and knob-operated spring latch, mortise locks are mounted in a steel case and set into the door more deeply than a cylinder lock. The actual lock mechanism is then inside the door rather than inside the knob, so even if a thief manages to remove the knob, he can't unlatch the lock. Usually, mortise locks use a locking rigid deadbolt, a few inches above or below the latch. The bolt is activated with a key from the outside and a key or a thumbturn inside. Although mortise locks are considerably stronger than standard cylinder locks, they can be defeated. Some mortise locks have short bolts that can be jimmied open, and on some older models a skilled thief can use a skeleton key. As a result, mortise locks aren't as popular as they once were. Finally, because mortise locks are installed deeply into the wood of the door, they can be difficult to replace. In any case, your sense of security will be heightened if you augment your locking system with an additional deadbolt lock elsewhere on the door.

In marked contrast to the old mortise locks are the push-button combination locks. No keys are required for these, and the appropriate sequence of buttons unlocks a rim-mounted deadbolt or spring latch. If there are 10 numerals on the keypad and the code involves seven numbers, then there are about 10,000 possible combinations—and it's highly unlikely that a thief will happen to guess yours. On the other hand, you and everybody else with access to your house will have to memorize the code.

Windows and Window Locks

To almost any thief, the window presents the only alternative to the door. It's that simple. If you think your doors are as burglarproof as they're going to get, then the only things you need worry about are your windows—all of them. Unfortunately, windows present a wider variety of problems than doors do. Your garage windows are probably constructed differently from your basement windows, which open differently still from your bedroom windows. And don't be overconfident because your windows are small or open only a few inches. Teenaged thieves can be surprisingly thin and capable of squeezing through pretty small spaces.

First of all, repair broken windows immediately. A broken pane not only advertises that your home might have other vulnerable points but it serves as an invitation to a thief who might be easily deterred otherwise. Then, as you would with your doors, make sure your windows are well maintained and operable. The putty around the frames should be solid and the frames should fit snugly. Keep in mind that burglars know that innumerable homes contain windows that do not lock. Anybody with a screwdriver can pry one open. If that applies to your home, it's time to do something about it.

Your windows can send out other signals as well, particularly if you install protective metal bars. A window-shopping criminal will think more than twice about taking a crack at a barred basement window. The same applies to grilles mounted on the exterior of your windows. Although its aesthetic value is minimal, securely fastened grillework lets you leave your windows wide open for ventilation and still be fairly confident that nobody can get in. But beware of letting the grilles and bars turn your home into a prison. Remember that windows are also escape routes in case of emergency. So any bars you install must be collapsible or easily removable from the inside. Grilles should have a quick-release lever that operates from the inside and lets you swing the grille outward.

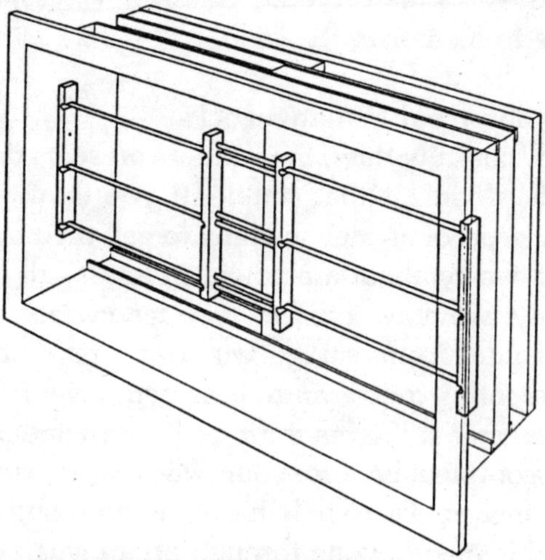

WINDOW SECURITY GUARDS

Steel bars installed on the inside of vulnerable basement windows make forced entry difficult and act as a deterrent against break-ins, but they must be designed to permit quick exit in an emergency such as fire.

You may want to consider replacing the glass windowpanes with stronger materials. After all, the reason thieves often enter through windows is because they're breakable. Break-and-enter artists have been known to carry tape with them so they can cover the window with it and quietly break the glass without shattering it. However, replacing the glass with Plexiglas, polycarbonates such as Lexan or special security film can have drawbacks, besides the extra cost. Plexiglas can yellow over time and is more prone to scratching. Furthermore, there might come a time when you or a firefighter has to break the window in a hurry. So regular glass does have advantages. The basement windows, at least, warrant window bars and strengthened or glazed glass.

The type of window lock you need depends largely on the type of window — sliding, double-hung or casement. Take a walk through a good hardware store and examine the wide variety of locks now available, and choose the ones that best suit your needs. There are also a few homemade systems that will further reinforce your windows. During all your deliberations, keep in mind that windows are sometimes used for emergency escapes. So if the lock you install needs a key or other removable part in order to open, leave that key, pin, handle or whatever in a designated and handy place.

The most common type of window — although not used as much in newer homes — is the double-hung window, which opens by sliding either up and down or sideways. Perhaps the most popular locking mechanism for it is the butterfly latch — to hold the window closed, you twist a butterfly-shaped metal attachment on the window frame into a slot on the sill. The butterfly latch, if it is not fitted with a key-locking mechanism, is extremely simple to open. If a burglar, even a novice one, looked through your window and saw that it was secured with a butterfly latch, he would know he could simply insert a knife blade or screwdriver to open it. Butterfly latches can be replaced with modern mechanisms that work much the same way except that they pull the window and frame tightly together, decreasing the chances that a thief could insert a knife or screwdriver. You can also purchase a key-locking latch lock, which prevents a thief from sliding the window open even if he smashed the glass to reach inside.

If you want more versatility, wedge locks or ventilation locks are specially designed to let you keep your windows locked in place but slightly open, so they can provide ventilation. These locks use keys that you should remove when you are not at home.

Windows that slide open can also be protected in much the same way as patio doors, with pins through the frame and locking bars. The trick, albeit very simple, is extremely effective, more so if you use more than one hole.

Since the 1970s, more and more houses have featured casement windows. Generally, these windows open outward and are operated by an interior crank. Though these sturdily manufactured windows are usually more securely constructed than their double-hung counterparts, they are far from impenetrable. For one thing, they too can be broken. And regardless of how they open, if a thief can cut through the glass and reach the lever or crank, the window might as well have been left wide open for him. Your best bet is to get rid of the old crank and buy one with a key-locking mechanism. If you would rather keep the old crank, remove it when you're not home so a thief can't operate it.

If you have a skylight, make sure it is securely installed and constructed of extra-strength material. You probably won't be using the skylight as an emergency exit, so it's suited for grillework or bars. And if your skylight can be opened, make sure it can also be locked—and keep it that way.

Window-mounted air conditioners present yet another problem. There have been reports of ambitious thieves roaming neighborhoods in a van marked "Air Conditioning Repair." Once they establish that a house is vacant, they rip out the window-mounted appliance and use the hole as a door. To avoid such a tragedy, secure your air conditioner with metal straps that are permanently fixed to the exterior wall. Or you can screw metal loops onto the control-panel side of the air conditioner and install a removable bar that passes through the loops and extends beyond the edges of the window frame, keeping the air conditioner where it belongs. Even so, window-mounted air conditioners will always be extremely vulnerable.

And don't forget your garage. Make sure its windows are secured. For all the sightseeing you do out of garage windows, you may as well apply glazing, Profilon or bars.

With today's technology, there's little reason for your home to be vulnerable. In many communities, you can ask a representative from the local police force to advise you on the best ways to secure your particular doors, windows and locks. And don't feel that you are bothering them. The police would far rather advise you now than answer your call after your home has been robbed.

CHAPTER THREE

FROM GOOD FENCES TO GOOD NEIGHBORS

"Knowing the neighbors is important, and my neighbors are great, but I can't expect them to guard my property. Another irony is, I didn't let them know I was going away for the weekend. I told them when I was going away for a couple of weeks, but not just a weekend. So on Labour Day weekend, as it turns out, my next-door neighbor heard the guy who had broken into my house. And she actually saw him leave. But she thought it was me."

In 1982, business was bleak for Carol Johnson. She was a fitness instructor in a large Canadian city where the fitness wave was just about to swell. But her students seemed to be losing interest. They weren't showing up for class. Curious, Carol asked a few questions and soon discovered that it wasn't her rigorous exercise regimen keeping them away from the gym, it was fear. There had been a few well-publicized rapes in the area, and her female pupils were doing their utmost to stay off the streets. Carol's next move had perhaps as much to do with physical well-being as do jogging or aerobics. Carol asked the police to hold a public meeting for any women concerned about safety on the streets. Afterward, armed with a little confidence and knowledge, the women started going back to class. And the incident marked the first of hundreds of occasions on which Carol Johnson, who would go on to develop one of Canada's foremost Neighbourhood Watch programs, would ask police to help residents of her community beef up their security and improve their quality of life. The absenteeism from Carol Johnson's fitness class planted the seeds for a home-security program that could serve as a model for anyone interested in improving the security of their communities.

Only six years after the fitness truants met the police, Johnson oversaw a Neighbourhood Watch program that included about 100,000 homes, 30 apartment buildings and 150 Neighbourhood Watch committees. Throughout the region, neighbors show their concern for one another by posting Neighbourhood Watch signs, holding regular community meetings and celebrations and participating in a project that brings the small-town feeling back to the big-city environment. It is a lot of demanding work, but the statistics bear out Johnson's efforts. Throughout Metropolitan Toronto, for example, statistics show that one in every 38 houses would be broken into in 1988. In the Metropolitan Toronto community of North York, that figure was more like one in 89. According to those figures, the area served by Johnson's Neighbourhood Watch program was three times safer than Toronto at large.

Later in this chapter, the development of a Neighbourhood Watch program will be explained, but for now, there is a parallel to be drawn between home security and protection in general. At the very heart of an effective Neighbourhood Watch program are two elements that blend together in the recipe for home security—a sense of community, and vigilance.

The first important point is that neighbors get to know and to care about one another. A Neighbourhood Watch program brings small-town neighborliness, where everybody knows everybody else, to an urban community. And vigilant people

will automatically call for help when a neighbor is in trouble. Those two traits are key elements of home-security procedures. Break-and-enter rates are higher in extremely dense urban neighborhoods, where people are unlikely to know who lives next door; and break-and-entries would never occur if the thieves suspected they were being watched by someone who was ready to phone the police. The more ways you can persuade a thief that he is going to be detected, the better.

There are a number of steps you can take to persuade potential burglars that someone's watching your property and that the chances of getting caught are quite high.

Fencing Partners

A good place to start is out on the street, from the other side of the property line. If good fences make good neighbors, they also send out good signals. Regardless of whether your property is surrounded by a short picket fence or a chain-link with locking gate, a strong, well-maintained, sturdy-looking painted fence tells prospective thieves that your property won't be an easy mark. Likewise, if your gate is rusting off its hinges or your fence has pickets kicked in, a thief will assume that the doors and windows of your house are probably not much stronger.

Ironically, there are some fences that thieves love. For example, if your yard is surrounded by a high wooden enclosure that blocks the view from the street, you might already have done the burglar a favor. After he is over the fence and into your yard, he won't have to worry about being seen. So, while a high fence may appear to offer protection because it is more difficult to climb, it may be more of a curse than a blessing. High fences can also block light from streetlamps. And a fence's shadow can provide a thief with a hiding place. A chain-link or open fence offers good protection while giving passersby (in particular, police patrols) the chance to see anything out of the ordinary happening on your property.

Many neighborhoods have bylaws that put limits on the height of fences. But even if your area has no such restrictions, be a considerate neighbor. To really nourish that small-town attitude in your area, keep the rest of the street in mind when you decide to build a fence. The folks next door aren't going to think much of anyone who builds a Berlin Wall right in the middle of a suburban development. You never know when the friendliness of your neighbors will pay big dividends in your efforts to secure your home.

They say the gateway to a person's house speaks volumes about the person. Your fence gate should be latched if not locked, and there's no real trick to installing a buzzer system that sounds inside the house every time your gate is opened. Also near the gate you might want to display a street address. Large, well-lit numbers on both the back and front of the house will help police or neighbors identify your home quickly. And your legitimate visitors and fast-food delivery people will love you for it!

If you think having the family name emblazoned on a

plaque out near the street lends an air of sophistication to your home, think again. A burglar can walk by, see your name and address, look up your number in the phone book, find out if anybody's home, and head in to rip you off. If you live in an apartment complex where you have to list your name near the door, don't identify yourself too clearly, particularly if you are a woman living alone. Your surname, maybe with a first initial, will suffice. The idea is to make a thief's work harder. That means not leaving notes to kids, delivery people or anyone else on the door or the mailbox. Imagine a burglar's delight when he reads that you'll be gone for a couple of hours.

What you want to do is discourage thieves. That takes lighting, property marking, maintaining a lived-in look, advertising to the burglar that your home is well protected, advertising to the burglar that your neighbors keep an eye on one another, and making enough noise in your home to convince the burglar that there's somebody or something (a dog, for instance) indoors.

Even though more and more burglaries are taking place during daylight hours, proper lighting is an important ally in your home-protection efforts. Indoor lights mean somebody is home, and exterior lights deprive a burglar of privacy. When you use lighting as protection, you have to do it carefully. For one thing, don't leave outdoor lights on during the daytime. Nobody does that normally, and it signals that nobody's home. And if you keep lights on inside the house, change your pattern. A thief watching houses in your neighborhood soon twigs to the same light that goes on and off at the same times, or the kitchen light that's always on when the rest of the house is in darkness.

Indoor lights should come on and go off with the same amount of irregularity that comes from having people go about their normal activities at home. So that means — and this is particularly important when you are gone from the house — you should install and use light timers. Timers are very inexpensive items, especially when their cost is compared to what you might lose if a thief thinks your dark house is a target. Most light timers are simple electro-mechanical devices, but you might also consider the more sophisticated programmable electronic versions. (You should also partially open the drapes or shades, to add to the lived-in look.)

Most timers are simply circuit-breakers that can be plugged into any electrical line. Install them in normally used rooms—the kitchen, the family room, and of course the bathroom. Be inventive and add a reasonably loud transistor radio to your timed lights, so the music and voices provide even more of a busy feeling to your home. Don't give in to the temptation to use television as a noise-maker. The fire hazard is too high. A radio and lights do the security job just fine. Once again, remember the neighbors. There's nothing more irritating for someone on the other half of a duplex or apartment than hearing the stereo next door start up in the middle of the night, especially when they know nobody is home.

Outside illumination is just as important as interior lighting. First, nobody wants to let a burglar have the advantage of darkness in the doorway so he can happily commit his crime unnoticed. So the first place you should install adequate lighting is in the entranceways. Ideally, all outside lights, wherever you put them, should be protected with mesh or unbreakable plastic and be located at least 10 feet (3 m) from the ground. Even low-wattage bulbs provide lots of security, providing they are protected and that they are not hidden by overgrown bushes, shrubs, stacked furniture, backyard equipment or other artifacts of regular family life.

If you want more elaborate lighting, one option is to literally bathe your home in light, virtually eliminating a thief's chances of working in concealment. Floodlights should always be aimed at your own house and not out toward the rest of the neighborhood, so like it or not, light will always be pouring in your windows. For a more intermittent effect, you could install your exterior lights on light timers. If you're the kind of person who forgets to program your timer, get photo-electric switches that will turn the lights on when the sun sets and off at dawn. Similarly, many security-conscious homeowners are opting for motion-sensitive switches that turn the exterior lights on only when someone comes near the house. Such a system lets your house sit unobtrusively in darkness until an intruder — or a visitor — comes by.

The ultimate illuminated property would have lights installed on the eaves, all the corners, outbuildings, fence posts, clothesline poles and telephone poles. Under ideal conditions, the property would also take full advantage of nearby street lighting, the garage light would augment the yard light, and neighbors would have a system of signaling each other with specially colored lights in case of emergencies. The outside lights would be controlled from the main house via underground wires that run through an unbreakable, water-tight pipe. Switches would be inside the house, near a window so the homeowner could observe whatever was going on outside (but not so near that a thief could break the window and switch the light off).

An auxiliary power supply would be installed in case the power goes off. Such optimum lighting conditions would significantly decrease the chances of that house being hit by thieves, all right. As long as the criminals tried to strike at night. Otherwise . . .

Alarms That Bark

For the thief who prefers the night shift, the task of casing your home to make sure nobody is there is more complicated than for a thief working days. For in the light of day, what is to stop a prospective thief from walking up to your front door and knocking? If someone answers, the thief simply invents a reason for knocking and goes safely on his way. However, if nobody answers the knock, and the thief knocks even louder and decides nobody is home, the casing job is done. The onus is on you to take steps that discourage the thief from advancing farther.

There's one sure-fire deterrent — a big professionally trained guard dog. To a suburban daytime break-and-enter artist, there's almost nothing worth the risk of coming face to face with a jumping, snarling, barking Doberman pinscher with teeth bared and eyes ablaze. What's more, you don't have to own a Doberman. A professional dog trainer will tell you that even an old-fashioned North American Bitsa ("Bitsa this and bitsa that") can be trained for household guard duties.

Whatever the breed, the dog you select as a home-security tool must be trained. You can't count on your untrained hound to scare off burglars. A well-prepared thief, with a chunk of meat, can turn your suspicious terrier into a quiet, friendly ball of fur. A more ruthless burglar might seriously harm a household pet. Don't count on Rover to keep your home safe. You might be risking the poor animal's life.

It's a different story with a trained guard dog. If you think you have sufficient room, enough energy and the serious need for a home-security device as radical as a guard dog, you can rest assured it will be an effective deterrent to burglars. But be sure you get extensive references before selecting a trainer. A good pro can customize your dog's training. Some dogs will attack when a stranger enters the guarded premises; others will only corner him until the master arrives. Included in a training program should be "re-fueling" every couple of years at the training school, just to make sure the bite is still as good as the bark. Depending on your chosen trainer, the cost of teaching the animal its duties might seem low relative to the price of an alarm system or the value of your home's contents. Of course, this does not include the lifelong costs of feeding, caring, walking, worrying about and cleaning up after your dog.

Consider your situation very carefully before you opt for a four-legged security system that barks. Yes, your deadly Doberman might keep the bad folks away, but lots of good folks won't like it either. Will a friend or relative be able to enter your house when you're not home, even if they are welcome? Both urban and rural homes receive more legitimate callers than illegal ones. You might be tempted to post a "Beware of Vicious Dog" sign, even though you have no such animal. Such a sign will deter more than just the burglars. There's nothing quite like a huge guard dog to lend a sinister atmosphere to a property. More importantly, if your dog even once bites the wrong person—a playful child, for instance—lives could be ruined.

Signed and Sealed

A thief might be discouraged if he sees a sign on your doors or windows indicating that your belongings have been specially engraved or marked to make them easily identifiable. In fact, some people think that the only protection they need is a sign stating that the house is protected by an electronic alarm system.

False advertising never works for long. So don't post the signs unless they are true. Entrusting your home security to a sign, in the fragile hope that the thief will read and heed, is like counting on your puppy to chase away a cat burglar. On the other hand, if you participate in a program such as Operation Identification, in which you use a special engraving tool borrowed from the police to mark your goods for easy identification, you will find the local police most encouraging and cooperative. (The police will often lend the engraver to you even if you haven't joined the program.) In some areas, the police provide a special marking pen with invisible ink that can be seen only under an ultraviolet light. Both systems are effective. A thief doesn't want, say, your VCR so he can watch videos. A thief wants a VCR so he can get money for it. And it is much more difficult to sell a stolen item when it is identifiable. After all, should an investigator come across the stolen item, it is the person who has it in his possession who gets charged, never mind whether he is the one who stole it.

Operation Identification, or some similar program such as Operation Provident, has several benefits: the stickers on your doors and windows discourage a thief, and your goods might be more easily recovered if they are stolen. Doubtful? Just ask the people at Portage La Prairie in Manitoba if Operation Identification is effective. Within one year of the launch of their program, the number of residential break-ins declined by about 50 percent. Operation Provident, which is a similar program for businesses, led to a comparable reduction.

Even if you don't join a local Operation Identification, you still ought to mark your belongings and keep a detailed record of all your household goods. If more people marked their property, the rate of recovery would be much higher than it is. Currently, about 80 percent of the stolen property that police find cannot be traced back to its original owner. When this happens, it goes up for sale at a public auction. And when you do mark your property, don't do it on a part that can be easily removed, such as the tone arm of a record player or the remote control of a VCR.

Operation Identification only goes so far. For instance, it is difficult to mark some items, such as delicate gold chains or gemstones. Indeed, the most valuable items in your home might be unsuitable for Operation Identification. If your warning sign isn't enough to deter the burglars, there are still some steps to take to protect yourself.

Safe and Sound

For jewelry that you seldom wear, the best bet is a safety deposit box, where it is away from your home and safe in the bank vault. When it comes to safekeeping for jewelry that does get worn, use your imagination. Build an out-of-the-way hiding place, and use it.

First of all, you should have a small collection of costume jewelry — glass and plastic baubles — that you can store in your bedroom drawers or in a small, obvious jewel box. These will be the thief's first targets, and if he is in a hurry, he won't be able to tell the difference between your trinkets and the real things.

You can't really have two sets of all your unengraved valuables, and you have to put them somewhere. These items include things like coin or stamp collections or photographic equipment, which you don't want to store in a safety deposit box because you use it fairly regularly. For these high-value portable items, you should choose either a strong room or a burglar-resistant safe. Converting a walk-in cupboard to a strong room is no mean task, but it is worth considering, particularly if your house is isolated or if you're frequently away from home. The walls and ceiling of a strong room should be lined with fire-resistant material, the door should be solid wood or metal with a reinforced frame and locked securely with the most modern locks. A strong room can serve a wide variety of purposes in your everyday activities, but it is important that you never forget its original purpose. *Always lock it.*

When you build a strong room, you are actually building your own equivalent of a bank safe. You might even install a combination or timed lock on it, making it even more like a bank's strong room. But there are other kinds of safes that warrant attention. While no safe is absolutely burglarproof, some are harder to penetrate than others. And like most things, effectiveness comes at a price.

The organization that tests the strength of safes, Underwriters' Laboratories of Canada, employs a variety of techniques to try to break into safes before grading them. A Class-2 safe, for example, has been proven to be tool-resistant for fifteen minutes. This means ULC testers used any tools available to crack the safe, and it took them a full 15 minutes of working time, rest periods not included. Compare the testing circumstances to those of a hurried screwdriver-equipped burglar, and it is clear that a Class-2 safe will present quite an obstacle to the average thief. A Class-3 safe is stronger yet. The best safes are constructed of drill-resistant steel and carry a ULC sticker. Safes can be bolted into a floor or remain portable, depending on your needs, and they can be purchased with either a combination lock, a key lock or a push-button lock. If you do purchase a combination-lock safe, establish a new combination as soon as you have the safe installed. Safes can also be incorporated into an electronic home-security system.

Be extra careful not to mistake a fire-resistant safe for a burglar-resistant safe. (Fire-resistant safes are described in Chapter Seven.) You can buy a combination of the two, but a safe that is built solely to keep your valuables safe from flames won't do much to keep them out of a burglar's hands. Finally, the biggest drawback to a safe is that in most cases it must be small enough that it doesn't take up too much room, and so you cannot put everything you own into one. Jewelry, collections and documents are fine, but you won't likely be fitting your VCR, TV or camera in there.

Good Neighbors

Most of the security equipment and aids — safes, dogs, lights, alarms, fences, locks and so on — that you either build or buy can be compared to the hardware of a computer. The software, however, is an inexpensive, easily learned program. It's called "community action."

Consider, for instance, Neighbourhood Watch. Neighbourhood Watch works not only to deter crime and catch criminals but it draws together neighborhoods, urban people who might not otherwise meet or deal with one another. At its heart, Neighbourhood Watch simply enables people to learn and care about one another. In doing so, they will also show concern for each other's property. Participants in the program post signs letting anyone entering the neighborhood know that if they do anything suspicious, their chances of being observed — and stopped — are high. In sharp contrast to the stereotypical don't-want-to-get-involved urban resident, Neighbourhood Watch members *want* to be involved. They're not afraid to report unusual occurrences to the police or to each other, so everyone can be on the lookout. To that end, an effective Neighbourhood Watch program relies on dedicated volunteers who are willing to stay in touch with their neighbors, either through newsletters, community events or even, as in one Ontario community, through a jointly sponsored telephone system called Rural Watch, on which participants can report sightings or be advised of suspicious activities for which to be on the lookout. Neighbourhood Watch can be an effective program. Ask your police department for details. At the very least, even if you don't become involved in an official program, take a lead from the Neighbourhood Watch people: report crime and unusual occurrences when you see them happening. Police will appreciate the help, and it's a sign of a caring community.

Another community program that encourages cooperation is Crime Stoppers. It was launched in New Mexico in 1976 by a policeman who guessed that two of the reasons many crimes went unsolved were: people were afraid that if they reported crime they would have to testify in court and perhaps face the revenge of the criminal; and people were somewhat apathetic but would respond to a reward system. The answer seemed to lie in a system within which the police, working with the media, would appeal to residents for help in tracking down criminals. Accounts of local crimes and descriptions of criminals would appear in newspapers and on television. A reward, as well as complete anonymity, would be offered. The program works, too. Since that first New Mexico project, Crime Stoppers has circled the globe. Around the world by the end of 1986, nearly 140,000 cases had been solved through Crime Stoppers. More than 12,000 crimes in Canada have been solved through the program. And according to one detachment in Winnipeg, a surprising number of people who helped the police didn't even accept the reward money. (That reward money must come from somewhere, and residents who would like to participate should band together and make some suggestions.) Perhaps one of Crime Stoppers' most significant characteristics is that it provides, once again, proof that a community working together can fight crime.

Another police-sponsored program, this time uniting police forces with house builders, is called the Shield of Confidence. Introduced in Hamilton, Ontario, in 1982, the Shield of Confidence is gaining a reputation and spreading across Canada. Police advise builders of new homes on how all the construction elements — the doors, windows, locks, lighting and identification — must meet certain standards to qualify for the Shield of Confidence to keep thieves at bay. The security standards are rigorous, but no more so than those described in the previous chapters on securing your home. The advantage of your home being made as secure as possible as it is being built means that it will have the strongest door frames, doors and windows, a lock on the garage and a complete set of door locks, all completely installed before you even move in. The Shield of Confidence homes carry a sticker warning thieves that the secure homes are part of the program. And some insurance companies offer premium reductions for homes built under the program.

As the crime rate increases, so does the cost of fighting it. The government of Canada encourages community-action programs that work to reduce crime. There are numerous avenues open to anyone interested in getting some kind of program operating in their community. If you do elect to get involved, not only will the odds of being broken into decrease but you'll also feel better about your community.

CHAPTER FOUR

ALARMING NEWS

"Ultimately, I guess we all have a certain naiveté when we buy a house. My God, you pay all this money, you pay for 25 years. Isn't it absurd that you don't take the steps to protect it more carefully? So why not spend the money? I think I spent about $1,200 so far, and the alarm system as I understand it shouldn't be too expensive. Whatever it is, it won't come anywhere near what I lost."

To Psych a Thief

No matter how personally offended their victims are when they imagine criminals rifling their bedrooms, thieves really don't think twice about the people they rob. If you get broken into, don't take it personally, says Dr. Sandy Wiseman, a forensic psychologist. Often called upon by lawyers to determine why break-and-enter artists behave the way they do, Dr. Wiseman has gained considerable insight into the motivation for crime. For one thing, he says, break-and-enter artists don't do their dirty work for thrills. The less challenge, the better they like it. Given the choice between your house and the less secure one next door, if they feel that it will be easier to steal from the house next door, then that will be their target. Thieves also will not enter a home if they think there is a good likelihood that they will be caught. They don't weigh the costs and rewards, like the average citizen would do when thinking about, say, risking a parking ticket: "Well, it'll cost me $10 to park legally, whereas a parking ticket would be $20, but I might not get it, so . . ." If thieves thought that way, they wouldn't commit crimes. Finally, Dr. Wiseman says, if a situation looks as if it is going to be extremely unpleasant, burglars will be just like everybody else and avoid the unpleasantness.

Dr. Wiseman's words provide some sound advice applicable to home-security systems: thieves worry most about inconvenience, detection, capture and hassle. Your fortified locks, windows and doors will certainly be somewhat of an inconvenience to the average thief. But if the burglar does get past the hardware, there is only one efficient way to mix detection, hassle and possible capture into the home-security recipe, and that is an electronic alarm system.

Household alarms serve several purposes. First, just advertising on windows and doors that your home is protected by an electronic alarm is not enough to deter some thieves. Of greater significance is that alarms make lots of noise. Some systems turn lights on. Both scare a thief and summon help. Also, a detection system can help capture the culprit at the scene of the crime. That, of course, is the thief's worst fear.

Finally, and perhaps most convincing of all, studies have shown that homes equipped with centrally monitored alarm systems are 15 times less likely to be broken into.

Electronic security devices — alarms, lights, communication devices — range in price from less than $10 to more than $10,000. However, when it comes to security systems, the extremely low-priced items could end up costing you more than you think. The cheaper a system is, the more liable you are to face false alarms. In fact, nine out of ten household alarms are false. In some municipalities, police have established programs whereby homeowners responsible for repeated false alarms have to pay the costs of the police response. In other areas, the police won't even respond after an excessive number of false alarms.

Another inconvenience of a cheap alarm is that it cannot distinguish between a moving cat and a moving cat burglar. As with so many other things in life, alarm systems become more discriminating as they get more expensive. But generally you will pay less for a comprehensive household alarm system than you would for a reasonable stereo system, an amount that's considerably less than the value of your household goods, as Alan points out.

There are two types of alarms—local and monitored. A local alarm triggers a loud siren or bell inside and perhaps outside your house, alerting the burglar, the neighbors and anyone else nearby that someone has illegally entered your home. A monitored alarm can also sound within the house, but it has the added feature of being connected, generally by telephone or a specially installed line, to a third party who can call the police. That third party in most cases is a professional monitoring station.

Local alarms are comparatively inexpensive, relatively easy to install, and the sound they emit might scare away the burglar. The ancient Romans employed geese for this purpose, but at least geese had the additional feature of biting. But if you can scare away a burglar before he can steal anything, your only problem might be a broken window. However, because the vast majority of local non-monitored alarms are false, passersby often ignore the signal, leaving the burglar to walk calmly away, carrying whatever he can get away with. A capable burglar can probably disconnect a badly installed alarm. (If you do install an alarm, tell the neighbors about it and ask them to call the police when they hear it. In *any* effort to beef up home security, it's important to have the neighbors on your side.)

Monitored alarm systems are more effective than local alarms. The most basic monitored system consists of a simple automatic telephone dialer that, when triggered, will dial a pre-programmed number and send a taped message to a third party. (Incidentally, the police prohibit use of their emergency number this way.) Presumably, that third party will then phone the police. If there is no answer at the number, some units then dial a second — or third or fourth — choice. Such a system has definite drawbacks. Say, for example, you have set it to dial a relative's home in the event of a break-in. If your relative has an answering machine and isn't home, then your emergency message will be taken down after the beep and go nowhere. Or if there is nobody home at any of the prescribed homes, then you're out of luck.

At the other end of the monitored alarm spectrum is the system monitored by a 24-hour central alarm station, approved and listed by the Underwriters' Laboratories of Canada. These systems are more expensive but they reduce the probability of illegal entry by about half and burglary losses in centrally monitored homes are decreased substantially. There is enough evidence of the effectiveness of centrally monitored alarm systems that insurance companies often offer customers a decrease in their home-insurance premiums if they subscribe to a monitoring service.

Here's how they work. At a ULC-approved alarm control center sit a number of fully trained professionals, 24 hours a day. (You will note that most alarm companies use ULC-approved equipment. It is vital, however, that the station be ULC-listed and approved.) If an alarm comes in over the telephone lines from your residence, a station operator should first phone your home to make sure it's not a false alarm, and then call the police, and individuals designated by you. The central station can also monitor your home's heating, cooling and fire-protection systems — and even your food freezer — and let the appropriate people know if something has to be repaired or looked at quickly.

Whether you opt for a local or monitored alarm, there are some characteristics that no alarm system should be without. First of all, check all equipment for the ULC label. Although alarm professionals are generally extremely conscientious, there are very few standards for alarm system installations. This may change in the near future, but for now, ULC labels are an assurance of product quality, at least. Then, ensure that your system has a reliable back-up power supply so the alarm can be triggered even if there is a power outage. The system should allow for adjustable entry and exit delay times, so you can go in and out of your house without tripping the alarm yourself. With the more complex alarms, you can adjust the delay time for the doors you use most often but maintain full security and instant response at other doors and windows.

Your security system should be straightforward enough that everybody in the house can use it comfortably, and you must make sure that the company that sells it to you is willing and able to train all occupants in its operation. On the other hand, the alarm should not be too easy to switch off or reprogram, or else a burglar can figure out how to do that. Look for a control system that lets you enter — and change — your own access code. Alarms should have warranties, they should be discreet in appearance (all wiring should be hidden), and they should be versatile so you can adjust them for your different living situations. You don't want the kids to trip the alarm every time they come home from school. And sometimes you may want to give non-family members access to your home when you are away. Designed for that purpose are alarms that let you provide temporary access without revealing a master code. An alarm system should be controllable from a variety of locations. Nobody uses the front door exclusively. You might want to be able to control the alarm system from, for example, inside the garage. And what if you're in the house and you detect an intruder? You may consider placing panic buttons in several places throughout your home. In fact, you may want to provide portable panic buttons to carry into the back yard or family room.

Sometimes, it is preferable to have a light go on when the alarm is tripped. There will be occasions when you want the window "perimeter" alarms switched on but the areas ("traps") switched off, such as in the evening when you are home but not in bed. Ask your dealer about these and other features.

Basics

When considering your security system options, you will come across two common terms: perimeter protection and interior, or trap, protection.

If a person's home is his or her castle, the perimeter protection serves as an electronic moat. Perimeter devices are installed on the entranceways to a house — windows, doors and skylights. One contact is installed on the frame of the window or door. Another is secured to the window or door itself, and it completes a magnetic electronic circuit. A signal is triggered when the circuit is broken — that is, when the door or window is opened and the two contacts separate. Ideally, these should be "concealed" contacts, installed in a manner that makes them virtually invisible.

Other types of perimeter alarms are:

• thin metallic foil strips that, although not attractive, set the alarm off if someone smashes a window;

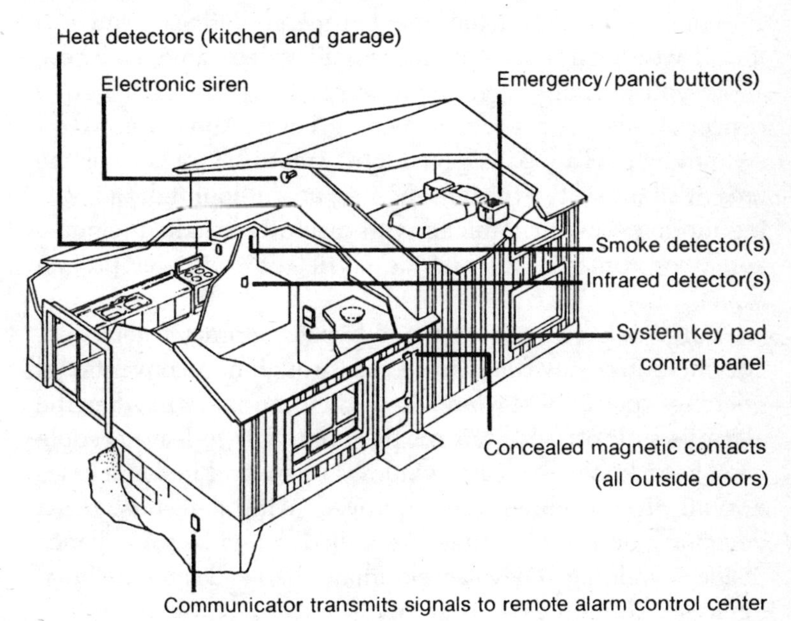

COMPONENTS OF A RESIDENTIAL ALARM SYSTEM

• glass-breakage sensors connected to the glass that react if the window is broken; and
• wired window screening, which acts as a regular screen but trips the alarm when it is cut.

Most burglars don't smash big windows for fear of the noise, and unfortunately many shock sensors can be activated by a passing truck, heavy wind or other loud noises. They're all sources of nuisance false alarms. The most effective glass-breakage sensor is the foil, often seen on store windows. But because it is unattractive, it is generally used for small basement windows that are concealed from the outside.

Another type of perimeter device, the outdoor motion detector, can either sound an alarm or switch on lights whenever someone comes within a prescribed area around your house. However, because a motion detector can't tell the difference between dogs and teenagers, it's advisable that it switch on only lights, not sirens or bells.

Finally, for top-of-the-line perimeter defense, you can install weather-resistant closed-circuit video cameras. These allow you to monitor your property on television, if you're concerned about prowlers or Peeping Toms. And if you have a swimming pool and kids, you might want to keep an eye on the area at all times. But for most homes, an outdoor light-activating motion sensor combined with carefully installed window and door contacts will provide sufficient perimeter protection.

Burglars have been known to bypass perimeter detection. For one thing, they can gain access through non-conventional openings such as attic windows, or they can sneak in when the alarm is shut off and wait for the occupants to leave. People have been known to leave windows open, and in some cases not all the windows are equipped with detectors. Area detectors, or traps as they are called, provide surveillance inside a building. They can eliminate the need for individual devices or foil on every window.

Area detectors are sensitive to infrared energy levels in rooms and corridors. Infrared sensors are either the new passive infrared type or the classic "electric eye" format.

Passive infrared detectors don't rely on any signal. Mounted inconspicuously on a wall, they detect infrared energy — a burglar's body heat, for example. With modern passive infrared units, an entire room can be monitored by one device the size of a cigarette package. The potential for false alarms is decreased with such a system because the temperature source — the burglar's body — must be moving to trigger the alarm.

The electric eye has two components, a photoelectric cell and an infrared light source. The source transmits an invisible infrared beam to the photoelectric cell. When the beam is interrupted, the alarm sounds. This type of infrared unit is suitable for narrow hallways or window areas, but it must be carefully installed. If the building shifts, as most do, the alignment will falter, the alarm will sound, and the device will have to be set up all over again. Set up the system along the route most likely to trap a burglar — although a skilled burglar looks for electric eyes and can usually avoid them.

One final type of trap alarm is the mat sensor, which goes underneath a piece of carpet or floor tile in an area where intruders are likely to walk. Mat sensors, when switched on, trigger an alarm when a certain amount of weight is placed on the surface. However, mats are limited in size and provide no guarantee of detecting an intruder. After all, what if a thief doesn't step on that required floorspace? They are best used in front of something you really want to protect, like a wall safe or piece of art.

Bear in mind that no single device provides complete protection. The web of protection afforded by infrared devices can be aimed high enough to catch a person crouching or standing while allowing a pet to walk through peacefully. But if your pet jumps onto the furniture, it may trigger the alarm. No system is completely foolproof. The system offering the greatest security is one in which several technologies work together. For instance, some trap alarms combine ultrasonic technology with passive infrared equipment, and the alarm won't be set off unless both detectors sense something out of the ordinary.

Installation

Admittedly, do-it-yourself household alarm systems appear to be simple to install. And they are, providing that the only elements you want to include are magnetic or mechanical perimeter devices on doors and windows, a small interior motion-detection device and a loud noisemaker inside the house. For several hundred dollars, you can buy a fairly sophisticated unit that controls a combination of perimeter switches, motion sensors and infrared detectors, as well as the necessary circuitry for later add-ons, such as smoke alarms or panic buttons. Most often, such do-it-yourself systems are described as "easy-to-install." Since they are battery-operated, their batteries must be replaced periodically.

Many modern homes are built with pre-wired alarm systems, making the home-installation even easier. If you are building a home, it's wise to install the necessary wiring just in case you decide to add an alarm system later. Otherwise, the home handyperson must figure out how to install the alarm-system wiring unobtrusively. As well, the siren, bell or horn on the exterior of your house should be well protected from the elements and not within reach of someone standing on the ground. The alarm should also be the kind that automatically kicks in if it is disconnected from its power source, as is the case if a burglar cuts the power line.

Homeowners who want to install a system can take tips from the pros. The Underwriters' Laboratories are in the process of establishing standards for three levels of protection for homes. A thoughtful alarm sales professional will determine which level your home needs and proceed from there. When designing a system, the expert will take into account a surprising number of factors — things you might not have otherwise considered.

The three levels of protection are high resistance, medium resistance and low-to-moderate resistance. At the low end, you should have sensors on all accessible doors and windows of your home. These sensors should be installed on a continuous closed loop connected to a master control cabinet somewhere on your premises. When the loop is interrupted, a circuit breaks and the alarm sounds. When ULC employ the term "accessible," they are quite specific. An accessible opening is any window, door or skylight that is within 18 feet (5.5 m) of the ground, 18 feet (5.5 m) from the roof of an adjoining building and within 14 feet (4.2 m) of opposite windows and 3 feet (1 m) of other openings or fire escapes. This level of security detects only people entering your home, not anyone who is already inside.

With medium resistance protection, interior motion detection devices, such as traps in strategic areas of the home, are added. With high-security installations, all doors, skylights, garages, basements, walls that adjoin other dwellings, and floors must be secured. Area detectors that survey all living spaces augment the perimeter detectors.

So you can install a security system based on professional standards yourself. However, because of the complexity involved, and because security companies own and operate their own alarm control centers, self-installation will limit you to a local alarm. Few if any monitoring companies will assume the responsibility of watching a home-installed system. And of course the insurance companies won't be offering you their vote of confidence in the form of a maximum premium discount.

Before installing your own system, conduct a physical checkup of your house, as outlined in the first chapter. Before sensors can work, doors may need aligning and windows may need firming up. Then check with the local police. There may be bylaws restricting the type of noise your alarm emits, and the police may already have established a system for dealing with false alarms. Then, check with your neighbors. If your alarm does not have an automatic timed shutoff, the people next door aren't going to appreciate your bells sounding all evening while you are away. Because community support is essential in all matters of home security, it is ten times as important that you have the cooperation of your neighbors when it comes to installing household alarms. Don't get them angry at you.

Almost nothing will irritate you, the neighbors, the police

or that yappy dog down the block as much as a false alarm, particularly if it sounds frequently. On the other hand, few events are as frustrating as coming home after a night out only to learn that your house has been burgled while your alarm system was not working. If you are considering installing your own system, and if you hate a false sense of security as much as you dislike false alarms, you have a lot of homework ahead of you.

The alternative is, of course, professional installation. You can start by inviting a security company representative to your home. They will be glad to oblige. Be sure to shop around. Of course you will be better off if you have a reputable firm conduct your security survey. To help you choose from among the many firms out there, ask friends, relatives and neighbors about alarm companies, and find out who installed your office system. Ask the company representatives for customer references. Inquire about warranties and after-sales service and consultation. Make sure their products and their monitoring stations are ULC-approved.

When a representative shows up for his or her appointment, ask to see up-to-date photographic ID. A professional thief would like nothing more than an opportunity to case your home in your company and find out precisely why you want an alarm system.

Above all, discuss all your needs and concerns with the representative. Let him know if you have a "revolving door" household, with people coming and going at all hours. Perhaps your family keeps a perfectly predictable pattern, day in and day out. Do you go away often? For how long? And finally, remembering that responsible alarm companies hold their clients in strict confidence, let the representative know what you're trying to protect. A respectable alarm expert will be honest with you and will not try to sell you something that surpasses your needs. If you live in a middle-class neighborhood and don't keep extremely valuable items on hand, your system probably doesn't need to be connected to the central station by a special "dedicated" telephone line. Your system can use the standard telephone line.

When the companies describe their central station services, ask for details. Are their stations ULC-listed and ULC-approved? Do their stations qualify you for an insurance premium reduction? Do they offer a key service, so that someone from the station can enter your locked home in case of emergency? If they can, the police and security company can react without disturbing you from your vacation, or wherever you are. In general, get as much information as you can.

If the representative is employed by a trustworthy company, he or she will be completely candid with you. The company's reputation is at stake, so you can expect honest answers about things like an installation schedule, the aesthetics of the system, the amount of wiring that will have to be done and how it will be covered up, and how easy the system is to use. The most common reasons for a false alarm with a professionally installed system are misuse and lack of proper care on the part of the users. So it is essential that everyone in the household learn how to use the system.

Depending on the security firm, you can either lease your system or buy it outright. Be careful to determine which arrangement you're getting into. You should also be able to find out in advance the total price for your system, materials and labor included. As with any major transaction, obtain an estimate in writing. (And read the fine print.) In addition to the cost of the hardware located in your home, central monitoring carries a monthly fee. If you have estimates done by more than one company, and if it appears that two alarm companies are trying to sell you what appears to be identical alarm equipment but at different prices, check their monthly fee. For a similar fee, different companies will provide different levels of service. Ask each company if the alarm signals are verified before the police are dispatched. This verification will greatly reduce the likelihood of false alarms. That's likely where they make up the difference. And remember that getting an estimate places you under no obligation to buy.

Although you can install some form of home-security system for under $100, industry experts estimate that a good system should cost no more than half of one or one percent of the value of your home and its contents. You should expect to pay more if you keep valuables in your home or if you are rich and/or famous. Then, providing you are satisfied with the estimate, the aesthetics and the service, the security firm should offer a one-week break-in period, so the people at the monitoring station can iron out any wrinkles in the system—that is, mold it to your family's lifestyle.

A typical professionally installed home alarm system might consist of:

- magnetic contacts on doors and windows;
- motion detection devices that account for pet movement (they're affectionately known as "pet alleys");
- smoke and heat detectors;
- strategically placed panic buttons;
- several household control panels; and
- a master control cabinet, which is usually located in the basement and doesn't have to be adjusted except when changes are made to the system's configuration.

The household control panels do the day-to-day job of arming or programming the system. On a good control keypad, family members are able to program the various sensors around the home so each zone is protected according to your wishes. If you are having a party upstairs, for instance, you probably want the basement windows and rooms armed and protected while the upstairs rooms remain unarmed. The keypads can also tell you, with lights or LEDs (light-emitting diodes), that everything is working or where an alarm has been triggered. When you return home, some control keypads will let you know if anything has affected the system in your absence. Control keypads should be tamperproof and small enough to fit unobtrusively into your decor. Many keypads have instructions conveniently located on the panel doors. If someone offers you a wireless alarm system, find out how long the batteries last and if other sources of electrical energy can interfere with the control system. Alarm effectiveness is always walking the fine line between economics, simplicity and the avoidance of costly false alarms.

The ability to monitor households is evolving daily. In the

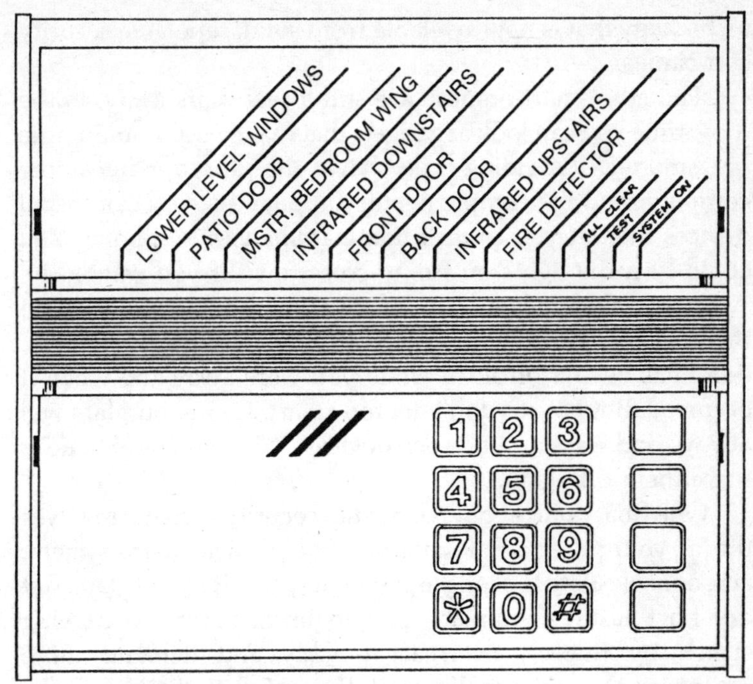

HOME ALARM SYSTEM OPERATING/COMMUNICATION PANEL
This panel is located near the door most frequently used for exit/entry. The burglar detection system is turned on/off by personal 4 digit code. Light bars beneath wording indicate the status of the system; for example, lights on illustration show that both patio and back doors are open.

future, we may see highly sophisticated "smart homes," in which all the appliances are connected to a central computer. In such homes, the burglar or fire alarm will not only trigger the necessary sirens but doors will close to isolate the fire or burglar. In the case of fire, these systems will switch on the lights that illuminate the safest exit. An electronic door lock, operable only through authorized pass codes, will keep a record of everyone who has gone in and out. A motion detector will alert parents when the children are close to a dangerous area, such as a fireplace or pool. The list continues.

Regardless of how advanced the "smart home" becomes, it is not going to render your current security-system obsolete. At the heart of even the smartest of homes is the same detection technology that is now available from solid, reputable security companies.

Though "smart homes" are still a few years away, in the meantime you can look at gadgets that make your home a little bit smarter. Providing you have the appropriate home computer and the complementary appliances, you can install devices that let you control home appliances by phone. You could turn the lights on from across town if you wanted to. Another clever gadget is an information module mounted on the wall that will literally tell you, in a synthesized voice, if there has been a break-in while you were away and when it occurred. But houses don't get too smart. Canny burglars will always find ways around such devices. What you have to do is scare them off again.

One final point about residential security systems involves timing your purchase. If you think you need an alarm system, act soon. Hesitate too long and you may be hit by burglars. But don't act hastily. Too many people, immediately upon being struck by thieves, overreact quickly and demand more protection than they really need. If you follow their example, you will never enjoy the peace of mind that an alarm system is designed to bring.

CHAPTER FIVE

A STATE OF MIND

"I'll tell you how skeptical I've become. The other day, I was home at lunch and there was a knock at the door. I looked through the hole in the door and there was a fairly well groomed young man, about twenty or thirty. Tight leather jacket, long hair, looked okay. First thing he says is hello, and then asks me for directions. You know, I'm pretty trusting and usually helpful. So I gave him directions and he left. But he was walking, and that seemed odd. I thought, 'That's not the normal way to ask for directions.' Most people wouldn't walk up to a stranger's house. You'd probably ask somebody on the street. So I think that guy was knocking on the door to see if anybody was home.

"I get more suspicious more often now. And there's an eerie feeling when you see guys hanging around who look like they might be vandals or thieves, and you get the feeling that they may be the ones who broke into your place. They might be the guys who went through your personal belongings."

The Pros and the Cons

The "needy kid" sting works like this: A youngster knocks on the door, preferably at the home of someone elderly. "I gotta phone my mom," the kid pleads to the homeowner. Moments later, the youngster's older partner shows up and offers to help. In fact, he's the one who volunteers to make the call. Then, with the partner waiting on the phone, supposedly for someone to answer, and all the while chatting with the homeowner, the youngster sneaks around quickly and steals whatever he or she can. If you think that's an unlikely scenario, just ask any of the people who fell victim to the trick recently. In one regular, middle-class residential neighborhood, before the police finally caught up with the culprits, the team netted more than $40,000 worth of money, jewelry and other items. The victims, all of them willing to help strangers, might have had secure locks and burglar alarms in place, but because they invited the burglars right into their homes, their security systems had no effect. (The best prevention here? Offer to make the phone call yourself on behalf of a "needy kid" or unexpected visitor — and leave them waiting outside.)

Personal security extends beyond break-ins. Not only is the unwitting homeowner liable to be a scam victim in his or her own home, but theft can wear many disguises. There are no fewer than 50 sections of the Canadian Criminal Code that deal with fraud. So, providing you are well prepared and alert to whatever problems might arise, the same peace of mind that comes from knowing your house is secure should accompany you wherever you go, into any situation. Whether you are on vacation, at the cottage, in a restaurant or only down the street, there are precautions you can take to make sure you don't become a victim. If you are a senior citizen, frequent traveler, parent, child or living alone, there are quite a few extra personal-security elements to take into consideration. The same applies to fire hazards. If you're not careful, a fire can completely destroy your home — and destroy lives.

Senior citizens make easier targets, perhaps because con artists count on them being more trusting and less of a physical threat. In fact, all the victims of the "needy kid" sting were elderly. Other schemes to defraud seniors have been attempted so many times that they are almost legendary. Most famous, perhaps, is the phony bank-inspector scam. The con artist phones a senior, tells him or her that he is a bank manager investigating the bank staff. The senior citizen is asked to withdraw money, at a specific time, from his or her bank account. If the victim complies, the phony inspector hangs around the bank at the same time that the victim is there and chooses the right opportunity to steal the money, whether it be by mugging or burglary.

No Canadian bank would ever ask a customer to withdraw money. It simply doesn't happen. So if someone does present himself as a bank inspector, refuse any requests he may make and call the police and the bank immediately. In any situation where complete strangers ask you to withdraw money from your bank, or where a sales pitch offers a once-in-a-lifetime opportunity, refuse to participate.

Less than crystal clear are scams involving visitors to your home. For example, if repair or delivery people show up unexpectedly, be wary. Before you let them in, ask to see identification. And trust your instincts. If you harbor any doubts about their authenticity, ask them for the phone number of the office they represent, and phone it to check up on them *before* allowing them to enter your home. Remember, if a delivery person or public official has a legitimate reason to enter your home, she will have the proper identification and she won't mind if you verify it with head office. On the other hand, if a stranger appears at your door and identifies himself as, say, a chimney repairman "just in the neighborhood, and wondering if you would like a free estimate for repairs or cleaning," keep him outside until you are absolutely sure his company is reputable. If he is interested in conducting legal business, he won't mind being "put on hold" until you satisfy your curiosity with a couple of phone calls. The local Better Business Bureau or Chamber of Commerce will be glad to help you alleviate your worries.

Just as Neighbourhood Watch experts claim that senior citizens are good "watchers," the reverse might be true. Ruthless con artists can keep an eye on seniors — or anyone, for that matter — and learn their habits. Therefore, try not to establish a firm routine in your banking or shopping hours, particularly if you live alone. If you regularly receive cheques in the mail on certain days of the month, instead arrange to have them deposited directly into your bank account so you don't have to make the same trip at the same time each month. If you must carry large sums of money, ask a companion to accompany you. Moreover, keep close tabs on your personal financial habits. Don't accept third-party cheques, never sign cheques until you are in the bank and avoid putting your signature (which can be worth money) on any document that you are not 100 percent sure of.

It's advisable for seniors to make contact with their community by joining Neighbourhood Watch, senior citizen groups, church organizations or some other service club. If there's no such group in your area, form one. Make a deal with somebody to exchange daily phone calls, just to check on one another. The essence of security always lies in a strong sense of community.

Simple precautions, once they become habits, can save you extensive misery. When you are returning home, for instance, have your keys ready before you get to the door so you don't present a target while standing in front of your house searching through your pockets or purse. Similarly, look inside your car before you open the door. Even though you should always lock your car doors, someone could be hiding in the back seat. In the same way that thieves don't like breaking into brightly lit homes, you are less likely to be attacked if you travel on illuminated streets. Therefore, avoid dark streets and alleys and shortcuts across unlit parks, playgrounds and parking lots. If you do your banking at an automatic teller, do it during the day, when there are other people around. Even then, scan the area around the bank machine for suspicious-looking characters. You should keep your money in a wallet inside a secure pocket, so it is out of sight of pickpockets. (Back pants pockets are an especially easy target.) Purses should be clutched under the arm and not dangled. But if the worst does happen, and a thief tries to grab your purse, let it go. The risk of serious physical injury to you far outweighs the potential loss of money.

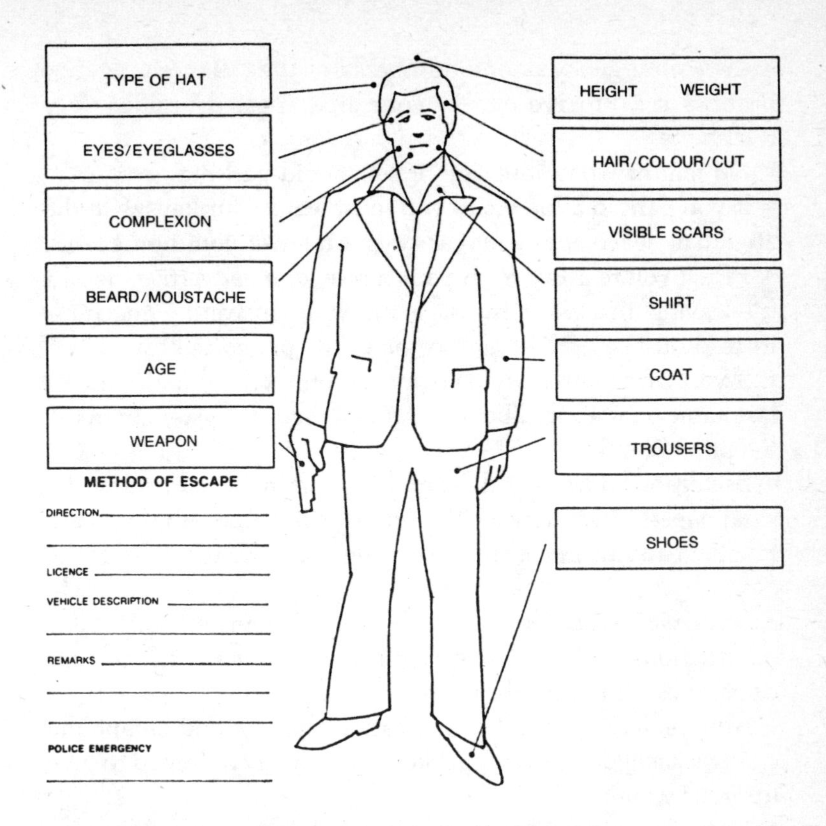

SUSPECT IDENTITY CHART

Learn how to identify people, just in case you are ever asked to describe a criminal or con artist. Regularly, after a person leaves a room that you are in, try to recall specific details about their appearance—their height, coloring, build, clothes, facial hair, approximate age, any noticeable characteristics. And as soon as you enter into a discussion with a stranger, make mental notes about his or her appearance. It's a skill that could save many other people from becoming crime statistics.

The hints for seniors apply to everyone, just as household security hints apply to homeowners, renters and apartment dwellers:

- Become involved in community groups like Neighbourhood Watch, Block Parents or service clubs.

• Keep emergency phone numbers—or the "911" emergency number if you have one in your area—posted beside every telephone.
• No matter what your age or sex, avoid parking garages.
• If you park in a lot where you must leave your keys with the attendant, leave only your car keys and never your home keys.
• When you're away from your house, don't advertise the fact by leaving that very message on your answering machine. Instead, just say, "We can't come to the phone right now."
• If you are a woman living alone use only your initials and/or last name on your mailbox, apartment security system or in the telephone book.
• Because weddings and funerals are often announced in the local paper, hire a house sitter (or ask a neighbor in) if you are involved and will be absent from the home for that part of the day.
• If you wear extravagant jewelry, don't do it in neighborhoods or situations where thieves might see it. Travel with your car doors locked, especially in cities.
• Always have extra transportation money and telephone change stashed in a small pocket in case you are forced to give up your wallet.
• Lastly but more importantly, get to know your neighbors.

Security on Wheels

One aspect of personal security that deserves serious attention is the car. Every day across Canada, more than 200 cars are broken into or stolen. In recent years, more than $200 million worth of cars were stolen. Many fall into the hands of professional theft rings who alter the vehicles and then resell them. Or else they are dismantled for parts. Other cars are stolen by amateur thieves who just want to go "joyriding." Professional thieves have a taste for quality and go after sportier, more expensive models. The most sought-after cars are Corvettes and Firebirds. Crafty car-crackers can enter a poorly secured vehicle in seconds.

In response to the escalating crime rates (and also the

increasing price of cars), electronics companies are forever improving car alarm systems, stereo equipment lock units, and sensing devices. But all of these have their limitations. So before you go shopping for an alarm system, consider your day-to-day driving habits. The best advice for avoiding either damage to the vehicle or total loss of it is:

- Park in well-lit, non-isolated areas, roll up the windows and lock the doors.
- Use your garage if you have one and lock it.
- Don't leave anything of value lying on the seat of your car.
- Never leave your vehicle registration slip or proof of insurance in the glove compartment.
- Don't "hide" a spare key somewhere in or on the car.
- Find out from a mechanic exactly how easy it would be to break into your car. If it's a snap, you can be sure that thieves will find that out too.

Car alarm systems are becoming much sought-after components, and as the demand increases, so does the sophistication of the units. Some alarm systems employ circuit switches that sound an alarm whenever the door, hood or trunk is illegally opened. These switches work like the little pins behind your car doors that make the interior light go on. Unless you are very handy with a drill and understand electronics, you should have them installed by a professional.

Another trigger often found on car alarms is the resonance sensor, which sets off the alarm in response to vibration from, say, a banging hammer or a breaking window. Depending on how elaborate your needs, resonance sensors can be inexpensive little battery-powered items that just sit inside the car or they can be more discriminating systems powered by your car's electrical system. Manufacturers have recently introduced ultrasonic sensors that "survey" a car's interior, sparking the alarm when they detect motion. The better systems employ a combination of sensors and switches. In any case, take care when choosing. Seek professional advice and, preferably, professional installation. You may nullify your warranty if you aren't careful.

Some questions to consider when selecting a car alarm system:

• Will the alarm function if the power supply is cut? It should. (But you won't want it to sound just because the car battery weakens, as it sometimes will in cold weather.)

• Will the alarm sound if the car is rocked by the wind of a passing truck or gently bumped by another car? It shouldn't.

• What kind of alarm does it use? Some honk the horn, which a thief might be able to disconnect, while others trigger a separate signal device. Some flash the lights. You can also buy a car alarm that lets you record a taped message. Imagine a thief's surprise if he hears a *voice* yell for help.

• In the more expensive category, some alarms send a signal to the owner's pager while leaving the thief to think he is working away undetected. When that happens, if everything works and you're absolutely sure that the alarm is reacting to a thief and not to an out-of-the-park baseball landing on the hood, you can phone the police and possibly catch the crook in the act. What's the range of the paging system? Will it reach beyond the confines of an underground parking garage?

• Does the car need an extra antenna to achieve that range? Many do.

• Does the electrical system for the alarm conform to that of your car?

• Is there a warranty?

• Will the alarm be set off by rapid drops in temperature?

• Is the alarm easy to program?

• Are the entry and exit delay times adjustable, so you will have time to get in and out of the car?

• What other features does your alarm system have? Remote-control units, with which you can unlock the doors, or set the alarm, are becoming popular. Car alarms can feature panic buttons in case of attack.

• Does the system rearm itself or must you remember to

switch it on every time you leave the vehicle? "Passive arming," so that you don't have to do anything, is more reliable. The novelty of having a car alarm eventually wears off, and you are sure to forget about turning it on every time you get out.

You will never, unfortunately, forget about it if the alarm screams like a banshee every time your car is touched. False alarms are numerous. The whole philosophy behind car alarms is to find that delicate balance of sensitivity while minimizing the chances of false alarms. Sometimes that sensitivity can be adjusted by either the owner or the manufacturer. Companies will claim that their systems can discriminate between a vandal scratching a car with a stick and the owner fumbling around in the dark trying to unlock the door. Greet such claims skeptically. The same applies to sales pitches that depend on the loudness of the alarms. A 60-decibel alarm is going to be just as effective as a 76-decibel noisemaker. And before you affix that little sticker to the window advertising that "This car is protected by so-and-so," think for a moment. The sticker may scare some off, but if your car is valuable and a pro thief really wants it, he probably knows how to disconnect the exact brand of alarm you are using.

Not all car criminals are out to take your car. In fact, owing to the general improvement in anti-theft devices, such as locking steering wheels and transmissions and fuel-line shutoff devices, Canada has seen a slight decline in the number of outright car thefts in recent years. However, there has been no similar decline in car looting. So even if your car has features to stop a thief from driving it away, and if you don't want to risk the potential nuisance that comes with false alarms, you can still fortify your car's valuable goods. You can purchase cassette players that won't work until you have punched in a secret code. Even if a thief gets his hands on that kind of machine, he can't use it. Or for only a few dollars you can use a cassette-locking device that is inserted into the tape cartridge. The device prohibits anyone without a special key from using the machine. You can buy a removable stereo system that mounts on an easy-to-use bracket, allowing you to take it with you whenever you get out of the car. (It doesn't mean thieves won't break in and trash your car looking for valuables, though.)

One final point. Every vehicle, whether it is a car, boat, RV or truck, should be equipped with a first-aid kit and a fire extinguisher. And never put fuel into a vehicle while the motor is running. On boats, have all passengers disembark before filling up with fuel.

Innocents Abroad

At certain times, you will be away from your neighbors, among complete strangers and in alien surroundings. There are few more devastating holiday events than robbery. So be careful on vacation. Before you leave, take all the necessary household precautions. The following tips apply whether you are going away for a weekend or for several months:

- Set light and radio timers throughout your house to give it a lived-in feel.
- Give an extra key to a trusted neighbor, who can visit once in a while and check your furnace, pipes and freezer and reset your timing devices. Let the neighbor know where you can be reached and when you expect to return. And be sure to let the neighbor know if you're coming home earlier or later than scheduled. Remember Alan: his neighbor actually heard the burglar in his house but thought it was Alan himself.
- If you are going on a longer trip, there are a couple of extra considerations. First of all, do you have the proper insurance in case of health problems or in case you have to cancel part of your trip? Before you purchase cancellation insurance, though, read the fine print studiously.

Arrange to have someone:

—pick up your mail and newspapers;
—mow the lawn or shovel the walk;
—park a car in the driveway once in a while; and
—generally keep an eye on things.

In some communities, you can hire bonded (legally safe) people to stay in your home or visit regularly while you are gone. Cancel the newspaper and arrange for the post office to hold your mail. Check that under your home insurance policy your property is protected while you are away. Some policies, in order to be effective, require that a homeowner's representative (relative or neighbor) inspect the premises occasionally. And most homeowner's insurance policies will insure against lost luggage or stolen jewelry, even when you're away from your home. Check with your agent to be sure.

Once you are actually on vacation, travel carefully. Travelers' cheques are safer than cash, but in many places, exchanging the cheques can be inconvenient. If you must carry a substantial sum of money, keep it close to you at all times, ideally in a secure money belt under your clothing. If you are staying in a hotel or motel, deposit cheques, airline tickets, passports, unneeded credit cards, valuables and surplus cash in the inn's safe. And keep a record of all the serial numbers of your travelers' cheques, credit cards and valuables in a place separate from the valuables themselves. Don't advertise your wealth. And avoid local "souvenir deals"—you might be buying stolen goods. Worse, you might be getting involved in a smuggling con.

You should not ignore fire safety when traveling, whether you are camping or staying in hotels. Campfires should be built downwind from your campsite. For extra safety, and particularly if there's a wind, build a trench or a small wall of earth or stones around the fire. Don't wear loose clothing near an open campfire. And keep some sort of extinguishing material nearby, even if it's only a bucket of water and a blanket. When you have put the fire out for the evening, rake the embers, douse them thoroughly, then cover them with sand or dirt. Barbeques, even the modern gas type, deserve equal caution. As much as possible, keep the hood open so unspent gas isn't trapped, and light the barbeque quickly. Anything less than extreme care in any situation involving fire could mean a fatality.

Hotels, too, deserve attention. Although commercial buildings are required to meet rigid building and safety standards, fires there can be deadly. After checking into a hotel or motel, plan an emergency escape route. This will take only a few minutes, and it will also give you a chance to find the requisite pop machines and ice dispensers. Take a quick walk down the hall and locate the exits. Count the doors and corners, because in a fire emergency you might not be able to see them. Look for fire alarm boxes and read the instructions on them. Throughout Canada, inns have emergency instructions posted on the inside of each room's door. Read these, too.

If the worst happens and fire breaks out, try to remain calm. If the fire is in your room, get out. Close the door and pull the fire alarm. Head for the nearest exit, and *don't use the elevators.* However, if you hear the alarm and you don't know whether it's for real or a false one, heed the directions on the back of the door. The alarm will indicate either "stand by" or "evacuate." If it is the latter, before you leave your room put your hand on the door. If you feel heat, leave it closed. If not, open it slowly, check the hallway, and, making sure you have your key, leave the room, close the door and head for the exits. Should all your exits be blocked by smoke or fire, go back to your room. Call either the front desk or the fire department and let them know you are there. If the phone isn't working, hang a sheet or other signaling device out the window or over the balcony and wait for help. If smoke starts coming under the doors, use wet towels or bedclothes to block it. Remember, most fire victims suffocate.

Vacations sometimes mean spending time at cottages. While a cottage seldom deserves the same extensive security system as a house, there are some things to consider. When you leave, use padlocks and chains to secure doors and windows as well as any boats or other loose items that you can't store right inside the cottage. Take valuables, firearms and liquor home with you. If possible, let nearby cottagers know what kind of schedule you will be keeping so they can be vigilant. You can offer to watch their place in return.

Any discussion of potential hazards can sound downright depressing. But most fire and theft precautions, after they become habits, consume only minutes of your time. This applies to your home, your car or your vacation. And once those simple precautions are followed, you can forget your worries and get on with having fun.

CHAPTER SIX

HOME IS WHERE THE HEARTH IS

"The day of the fire we had our woodstove burning. Of course smoke and soot builds up in the chimney—that can be extremely flammable. Some material stayed lit, landed on the roof, and the fire started. The thing is, because it was up on the roof, thc flames had a real head start before anyone inside knew what was happening.

"Somebody out in the yard ran in and yelled, so we had time to get the family out. We're out in the country and it took quite a while for the fire department to get there and we lost everything, except, thank God, our lives.

"Of course it's easy to say afterward how we should have taken care of that dirty chimney. It's always easy to figure out what you should have done after disaster has struck."

For centuries, fire has served mankind as a practical tool—for cooking, heating, energy, motor power—and its usefulness cannot be overstated. Fire has also worked as a symbol: a symbol of power and of terror. There are good reasons why some people worshipped flames while others saw them as the epitome of evil. Fire strikes indiscriminately, regardless of

CANADIAN FIRE LOSSES AND DEATHS

				Residential Properties Only	
Year	Total No. of Fires	$ Loss	Deaths	No. of Fires	Deaths
1983	70,953	815,967,016	539	34,517	488
1984	70,730	929,479,253	598	33,185	511
1985	70,061	925,955,586	550	33,150	469
1986	67,884	973,458,302	553	33,184	436
1987	67,168	956,207,142	516	30,735	439

Source: Fire Commissioner of Canada

- Residential properties account for almost 50% of all fires.
- On average 85% of all fire deaths are in residential fires.
- Sources of ignition (1987 typical). The primary identifiable causes of fire are:
  - smoker's materials (20%)
  - Electrical problems (17%)
  - Cooking equipment (13%)
  - Heating equipment (10%)

whether you are rich, poor or in between. More than burglary, more than vandalism and even more than financial loss, an unwanted fire can bring to the doorstep of cautious, law-abiding citizens more heartbreak and trauma than almost any single sad event. Fire destroys dreams. Fire kills.

As with personal and financial security, it is important to be concerned with fire safety, both at home and elsewhere. If people were a bit more cautious, if people took just a few extra precautionary steps, most of the tragedies that result from fires could be avoided. A fire-conscious householder should remember that even when the cause of the fire seems completely unpreventable, as in the case of a lightning strike, steps can be taken to minimize the damage and suffering.

Happily, all of society has been hard at work fighting fire

and decreasing its destructiveness. Over the past few decades, firefighters, builders, citizens and researchers involved in the ever-changing technology of fire control have worked together to keep fire losses from increasing. And their efforts have been paying off. Fire-prevention technology, enforcement of building codes, industrial safety standards, educational programs, improved materials, safer living habits and up-to-date emergency communication systems have greatly reduced fire-related tragedies. The number of fires, as well as the rate of deaths due to fires, has been decreasing. Over this past decade, about 70,000 unwanted fires burned annually in cities, towns and elsewhere across Canada. This figure is down from a decade earlier, when the average number was more like 80,000. In 1978, 844 Canadians died in fires. By 1987, that figure had dropped to 516. And that declining trend continues.

But the cost of fire control continues to rise, as does the monetary value of the property lost to flames. In 1978, there were just over $2 billion in insurance claims filed for property lost in fires. And even though the number of fires was lower, by 1987 claims had almost reached the $3-billion mark. So while fire-prevention efforts are paying off, we can't afford to sit back and let the professionals do the worrying. Besides, almost half of all fires strike one- or two-story residences, and 85 percent of fire deaths occur in these places, the places most Canadians call home. If your home is one hit by fire, or if a fire-related tragedy strikes one of your family or friends, you will agree that even one fire is too many.

The majority of accidental household fires can be traced to one of four sources: heating equipment, which includes furnaces, fireplaces, woodstoves and chimneys; kitchen and outdoor cooking equipment; appliances, and electric wiring; and basically anything that involves flame or fuel, such as smoking, burning candles or using kerosene.

With that in mind, do yourself a favor: refer to the Checklist on Fire Prevention in the Afterword. Then take a long hard look at those four aspects of your household, go over the fire-safety guidelines below, and make your home a whole lot safer.

Beating the Heat

As an example of space-age technology that's economical, effective and safe all at the same time, you need look no further than the modern furnace. But just like any other piece of modern machinery, the furnace is safe and efficient only as long as it is maintained and operated according to the manufacturer's specifications. Having the furnace technician visit once a year won't cause you nearly as much anxiety as, say, seeing your dentist regularly, and the furnace technician's visit may save your home from fire. Insist that the professional clean the furnace, check the safety controls and examine the chimney and flue for buildups and leaks.

Here are a few other heating-system tips for which you don't need a visit from a professional.

Furnaces and the basement in general should be kept clean, and you should take careful note of what's down there. Don't use the furnace area as a storage space for paints, painting materials or turpentines and solvents. In fact, you should not store any combustible materials in the basement, and preferably they should be kept out of the house altogether. Rags, discarded newspapers, solvents and cardboard not only add fuel to a fire; they can help start one. Throw oily rags or paint-smeared materials away. If you must keep paint or other flammables in the basement, store them in tightly sealed steel containers, away from the furnace.

The Insurance Bureau of Canada says that homeowners must verify that all heating plants, pipes and appliances are the minimum approved distance away from walls or flammable material. Check to see if the walls, floor and ceiling around your furnace are either built of fireproof material or coated with same. Likewise, make sure the pipe that connects the furnace to the chimney fits snugly, has not rusted and has no holes—holes through which deadly sparks might fly. (If repairs are necessary, or if you think your heating fuel is leaking, don't tackle the job yourself. Get a pro.)

One in ten fires start because of faults in the furnace, fireplace, chimney, flue, cooking stove or heating stove. When these integral features of your home are kept in good working order, clean and well maintained, you face a minimal risk of losing your home in a fire. But if your chimney is dirty or your heating stove improperly installed, you're asking for trouble. If you see any abnormalities around your furnace, such as cracks or loose mortar, attend to the problem immediately by contacting your home-heating company. In cottages, where fireplaces and woodstoves are common, take even greater care around fires than you would at home.

Most people who die in fires do not burn in the flames. They die from smoke inhalation or suffocation by poisonous gas such as carbon monoxide, which is odorless. What's more, there doesn't have to be a fire for a chimney to cause serious problems. A recent coroner's jury in the Ontario town of Picton decided that the deaths of six people were the result of carbon monoxide poisoning and extensive burns resulting from a faulty chimney.

This hazard should be of concern for several reasons. First, airtight homes can be prone to back-drafting, which draws air back down the chimney. And that air could be deadly. Modern airtight homes need careful ventilation, particularly if there is a fire blazing in the fireplace. Also, chimneys warrant careful maintenance whether you heat your home with a furnace fueled by oil, natural gas or propane or a woodstove/fireplace fired by wood. In the first case, because the modern furnace is thermostatically controlled and therefore doesn't stay on long enough to keep the chimney continuously hot, the smoke can condense and deteriorate the structure of the chimney. Also, because cold air is denser than warm air, a cooled-down chimney contains a column of cold air, which the furnace tries to push out every time it starts up. This can sometimes lead to a dangerous recycling through the furnace of the smoke and other combustion products. This can also happen if the chimney gets blocked by debris, a bird's nest or an unfortunate dead animal.

If you want to look up your chimney to make sure nothing is blocking it, hold a mirror inside the inspection or cleanout opening, or open the flue over your fireplace and hold a mirror under it. If you can see daylight at the top, you're probably okay. However, you can't tell if your furnace is recycling burnt material just by looking. If an oil furnace is doing so, you should be able to detect a strange odor. You can tell if a gas furnace is recycling the material if there is a lot of moisture spilling from the chimney's draft hood. Chimneys must be properly insulated. In a poorly insulated chimney, gases cool too quickly and condense, hurrying not only the erosion of the chimney materials but also the ensuing hazards.

Your chimney should be kept free from creosote — a flammable byproduct of combustible material — and it should be built so that it extends above the highest point of your roof. Any unused flue holes should be sealed with non-combustible material, not left open or merely covered with paper. Improperly installed flue pipes are the starting place of most fires with any wood-burning appliance. With masonry chimneys, make sure the collar around the flue pipe is metal or tile and securely anchored.

If you use a fireplace or woodstove to supplement your primary heating source, each heating source should have its own chimney. And as cozy as a fireplace or woodstove might seem, either one deserves extra caution in its use. First, make sure your fireplace is screened off to prevent sparks from entering the room. Try to keep the area around the front of the fireplace clear of flammable materials such as magazines, cushions or a woodpile. Again, that's an obvious but easily overlooked point, like remembering to lock the door when you take the garbage out.

For your own sake and that of the neighbors, don't burn paper, cardboard or other trash in your fireplace. It's not an incinerator, and glowing embers can travel a long way and wreak havoc to your property and other people's. The best material for burning in a living-room fireplace is dry hardwood. Second-best is softwood. After a little practice, you ought to be able to start a pretty good fire using about five small sticks of dry kindling and a few pages of the newspaper. Under no circumstances should you ever use combustible liquids like oil, gasoline, kerosene or barbeque starter.

Fireplace fires, like furnaces, should be cozy but essentially odorless. Even though your blazing fire might smell delightful, the odor is actually a sign that potentially dangerous chemicals are being released into the air. Burning wood, if not properly ventilated, emits carbon monoxide, carbon dioxide, formaldehyde, formic acid, carbolic acid and other irritants. So if you do smell smoke, call in an expert. Fireplace ventilation problems can be caused by a variety of things, including a blocked or filthy chimney, a closed damper or another exhaust system such as your dryer or range-hood fan operating at the same time that the fire is burning.

Woodstoves are extremely popular alternative-fuel home-heating devices. But if you don't use it properly, your woodstove can become a nightmare. Woodstoves should not be called upon to heat anything more than the room in which they are located. Before you use it, make sure your stove meets with Canadian Standards Association (CSA) approval. The CSA and fire marshals' offices across Canada have stringent guidelines for the installation of woodstoves, including the minimum distance needed between your stove and adjacent walls and furniture; the number of millimetres a flue pipe can be inserted into the chimney; recommendations regarding the design and construction materials of the stove, and so on. Should you be thinking about installing a woodstove, plan very carefully. Keep in mind that woodstoves are reminders of a time, not too long ago, when Canadians relied on them as the only source of heat—and that house fires occurred far more frequently in those days.

What's Cookin'?

There are two areas of your house where you use extreme heat daily: inside the furnace and in the kitchen. The firebox of a furnace is designed so that, if everything is going smoothly, a homeowner need never have anything to do with it. It will be maintained properly with an annual visit by a furnace repair technician. However, the kitchen is another story. There, inside the oven, around the toaster and on the stove, almost everyone is expected to work comfortably with extremely high temperatures and complicated electrical appliances. Fire safety in the kitchen depends on two equally weighted factors: proper appliance maintenance plus personal care.

As you consider your home-security and fire-safety needs, you invariably make one recurring observation: a tidy home has fewer problems. The same thing applies to your kitchen. A carefully organized kitchen area is less fire-prone than a disorganized one is. The kitchen should be kept clear of household cleaning supplies, which are often extremely flammable. This includes oily rags, cloths used for polishing furniture, and waxes.

On the other hand, don't put the materials too far away, because it's imperative that you keep your kitchen clean. The grease that builds up around stovetops is extremely incendiary. If you must keep cooking grease handy, store the containers away from the stove. About 16 percent of fatal household fires start in the kitchen, and if you have ever seen a grease fire spread, you will understand the link between cleanliness and safety only too well. Ovens, too, deserve regular cleaning, for cooking in a soiled oven can sometimes result in an explosion.

Don't use kitchen appliances for anything other than their original purposes. How often have you arrived home with an armload of groceries and piled some of the bags on top of the stove while you unpacked? What would happen if someone had accidentally left a burner on? When you're not using a pot or pan, take it off the stove. It's too easy to absentmindedly switch on the wrong burner, and on some older ranges, heat from the oven is sufficient to ignite something on top of the stove. As obvious as that precaution might seem, the most common type of stove fire is caused by somebody placing a flammable item directly on or right next to a hot stove.

Never use the oven as a heater. Don't hang damp clothes over the stove to dry, for example, because they might fall and start burning. And avoid using the toaster or stovetop burners to light candles, cigarettes or anything else. The wick or paper might leave a potentially deadly incendiary ash. Matches you keep in the kitchen — or anywhere, for that matter — should be kept well out of the reach of children, stored safely in a sturdy container, and be of the safety-match variety. And be sure not to toss them into wastepaper baskets when you're through. Douse them in water first. The same goes for cigarette butts.

Always turn pot handles away from the front edge of the stove so they can't be accidentally hit and knocked over; and don't walk away from a pre-heated pan or pot without first switching off the burner. If there is a flare-up in a frying pan, don't move it. If you do, the air current will fan the flames, leading to disaster for you and perhaps for your entire home. And don't pour water on it, as that will only spread the problem: not only will the blazing grease travel farther on the water but the water will turn to steam and splatter the grease even further.

Every kitchen should be equipped with some firefighting device, such as an extinguisher or fire blanket (these are discussed in the next chapter), but baking soda does work on small grease fires. The best way to extinguish a small fire on the stove is to put a lid over the pot or pan to contain the fire and switch off that burner. Although it's tempting to switch on the overhead fan to get rid of the smoke, don't do it. The flames and smoke will be pulled into the ventilation system, endangering the entire house. And make sure everybody in your household knows the right way to handle small kitchen fires.

The kitchen is also the spine of your home's electrical system. Consider the assortment of electrical appliances there. First there is the heavy-duty oven. Then you have the refrigerator, freezer, dishwasher, microwave, food processor, toaster, kettle, coffee maker, electric knife, garbage disposal, electric frying pan, lights, radio and whatever other gadgetry is available. For your own peace of mind and body, they should all be products approved by the CSA. The Canadian Standards Association represents industry, government, educational institutions, the scientific community and consumer organizations. If an appliance carries a CSA sticker, you can rest assured that the product has been tested for safety and performance and meets high standards. This applies to electrical cables, interior components and finished products. But it only applies to those products at the time of production. If you have altered them, or if they're old and worn, the certification no longer applies.

House Rules

Even though the kitchen is action-central for so much of a household's appliance usage, you should not ignore fire-alertness elsewhere in your home. About one-seventh of all fires are caused by the misuse of electricity—and not only in the kitchen.

Assess all your appliances in every room. Try to recall first of all whether you get shocks from any of them. If the only shocks you can think of are the type you get from shuffling around on the carpet in your socks, don't worry. But if any appliance or outlet emits shocks, have it checked by a professional electrician as soon as possible.

Additional electrical hazards to be on your guard against include:

• Homemade and non-CSA-approved fixtures and appliances;
• Loose joints and connections that might overheat;
• Deteriorated and frayed insulation on wires;
• Wires that have been chewed by pets;
• Overly bright light bulbs;
• Light bulb sockets used to operate appliances;
• Appliance plugs that are hot to the touch when plugged into the socket;
• Electrical cords that run under appliances or under carpets;
• Spliced cords;
• Three-pronged connectors into two-pronged receptacles;
• Extension cords used in place of permanent wiring;
• Hidden extension cords;
• Appliances used near sinks, bathtubs and other sources of water;
• Improperly rated fuses; and
• The infamous octopus outlets.

As dangerous as these situations are, it is amazingly easy for a householder to suddenly be faced with one or more of them. As a house becomes more efficient and modern, the more appliances appear. It's easy to overload your circuits.

Other electrical problems are sneakier than those listed above. During the course of everyday life, you don't busy yourself with the permanent and heavy-duty wiring of your home, yet it is among these more potent cables that other prospective fire hazards grow. Your permanent wiring should be inspected by an electrician at once if one of the following conditions arises:

- Regularly blowing fuses;
- Problems inside metal power boxes, such as loose joints or bad connections;
- Sparking from outlets or outlets that get warm;
- Badly strung wiring;
- Ungrounded switches and fixtures; and
- Any other signs of electrical problems, such as wavering current, surges or interference with televisions or radios.

Just as a forest fire can spread for miles underground without anyone knowing about it, a house fire can start among the permanent wiring and spread quickly behind the walls, undetected until it's too late. A room can be destroyed completely — furniture and everything — within seven minutes of a fire starting.

Fire can also start when you have an accident while using flames, heating units, gasoline or other flammable material. For example, perhaps you use electrical heating elements around the baseboards, or maybe you augment your cellar heating with baseboard heaters. If they are operable and CSA-approved, the heaters should work without posing any hazard. But if you place them too close to, say, curtains, you could be facing serious consequences.

The same logic applies to anything you do around the house that involves open flame. Almost 13,000 fires across Canada during 1987 were the result of misused open flame. Matches must be kept in a safe place. Never leave lit candles unattended, and keep them away from anything that could catch fire — curtains, lampshades, dried flowers on the dinner table. Make sure children or pets — or your own elbows — won't knock them over. Don't smoke in bed. Smoking represents about 20 percent of the known causes of fire in Canada, particularly when it is combined with heavy drinking. If anyone smokes in your house, make sure clean, big ashtrays are always nearby. But don't put ashtrays under or near curtains or drapes. And don't strike matches in closets, garages, near the furnace or anywhere else flammable material might be.

Be especially alert at those particular times, especially the holiday season, when decorations, candles and dried-out trees add to the fire potential in your home. In almost every case, electric candles are preferable to real ones, just as electric lights are eminently safer than kerosene lanterns.

In fact, when dealing with live flames or flammable liquids like kerosene, gasoline or naphtha, the possibility of tragedy is never far away. About one in every ten fires is caused by mishandling such deadly materials. So handle them sparingly and carefully. Don't even use fuel liquids as cleaning solvents. You can't tell where the vapor travels or where it will come into contact with a spark. Also, some flammable fluids vaporize and become explosive when exposed to air. Only use approved containers; only use them in well-ventilated areas; and only use flammable liquids with the utmost discretion.

All the warnings and all the insurance you can buy will amount to next to nothing if, some night, fire breaks out in your home and you or your family suffer. It's more important to secure your home from fire than from any other kind of hazard. So act now.

CHAPTER SEVEN

WHERE THERE'S SMOKE . . .

"Even though the fire was pretty well contained on the third story, the firemen really poured the water to her, so the whole house was a mess. Some of the things we lost were irreplaceable. But the worst, I think, was losing all our pictures, baby pictures, pictures of when we were younger. That's a lot of memories.

"Fortunately, everybody got away without getting hurt. But every time I think about the fire I think of my little boy and I just want to hug him. Thankfully, he's all right, but we'll never forget how close we came to losing him."

To anyone who has lost a relative or friend to fire, there's little consolation in knowing that most fires can be prevented. Preventing fires altogether is one thing. Reacting to a fire that has already started is quite another. It is at that stage, after something is burning or smoldering, that the homeowner has to rely on his or her defensive planning. The safest home will be well equipped with fire-control equipment, and the people living there will be well versed in what to do if a fire strikes.

Smoke Detectors

Unlike many of life's activities, equipping your home for fire safety is an extremely inexpensive activity. The prices of fire extinguishers, alarms, detectors and blankets vary, but they are designed and priced with the safety and well-being of Canadians in mind. Furthermore, when stacked up against the cost of replacing a home, or the loss of a human life, there's no comparison.

But even if not legally required, no home should be without at least one smoke alarm. Time and time again, fire departments, when reporting on a fire-related death, advise that the tragedy could have been averted if a working smoke alarm had been in place.

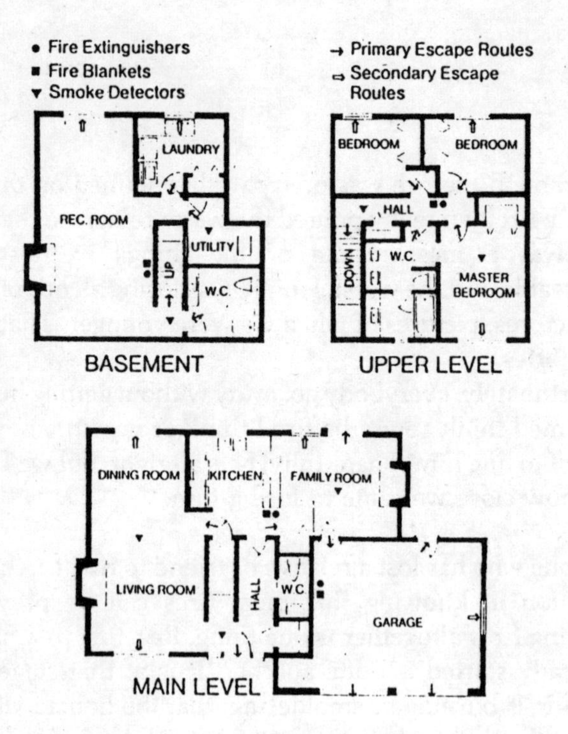

HOUSEHOLD FIRE PROTECTION

There are two types of smoke alarms—ionization and photoelectric. Both have the ability to detect smoke, to sound an alarm, and to provide effective, reliable early warning that a fire condition exists.

They can be wired into your home's electric circuitry, or be battery-operated. In homes with alarms connected to the electrical system, the addition of battery-powered units ensures that the detection capability continues even if power is cut off.

Ionization-type detectors use a tiny, innocuous amount of radioactive material which transforms the air inside the detection chamber into an electrical conductor. When smoke particles enter this chamber, the current is changed and the alarm is activated.

Photoelectric detectors incorporate a light source and a light-sensitive cell, called a photocell. Smoke entering the unit scatters and deflects the light, thus triggering the alarm.

While ionization detectors generally respond more quickly to flaming fires, and photoelectric to smoldering fires, both are equally effective in the home, with a reaction time difference that is negligible. Of course, if you are particularly concerned, you can install both types.

Most smoke alarms have small red indicator lights that tell you they are powered. With just a glance and a few minutes of your time, you can conduct an inspection "tour" once a week to make sure they are ready to function. At the same time, if your detectors are so equipped, you can push the "test" buttons to check the circuitry is functioning.

Smoke alarms have been around long enough for the authorities to have spotted a problem that has, on numerous occasions, resulted in tragedy. Battery-operated units won't work if the battery has run-down or has been removed. Some of us, apparently, are neglecting to replace worn-out batteries, or are removing them because of a few false alarms. Don't let this happen in your household; it simply isn't worth the risk. A smoke alarm will signal, by beep or flashing light, when the battery is getting low. Never ignore this warning. Replace the battery immediately.

The only other maintenance required is a periodic vacuum cleaning, which is particularly important if the unit is a few years old. Dust or grease build-up detracts from the detector's ability to sense smoke.

When you're considering your smoke-detector options, ask the salesperson about the model's susceptibility to false alarms (the cheaper it is, the more often you will have false alarms), its ease of installation, its volume (below 80 decibels is too quiet) and its warranty. Check that it is listed by the Underwriters' Laboratories of Canada (ULC). If it's not, don't buy it; it's illegal.

A smoke detector's prime purpose is to grab your family's attention before a fire gets out of hand. So be careful where you put the alarm. If you are determined to use only one in your home, put it between the bedrooms and the living areas, preferably in a hallway so that it picks up smoke coming from the living areas but wakes people in the bedrooms.

A single detector is really inadequate, though. An ideal home would have a detector in each room. If that is not feasible, you should at least have one on every floor. Bedrooms, corridors and the tops of stairwells are good locations. In a split-level home you should have two alarms, especially if the two levels are separated by a door.

Install the smoke alarms according to the manufacturer's instructions, no more than 5 feet (1.5 m) from the center of the area they should monitor. Whereas a large home merits several smoke alarms, smaller dwellings, such as mobile homes or cozy bungalows, can become suffocatingly thick with smoke in a matter of seconds, so it's absolutely essential to use high-quality detectors in them.

Be careful to set them more than 6 inches (15 cm) from the wall or they will be ineffective. If you cannot mount your detectors on the ceiling, affix them to the walls between 6 inches and a foot (15-30 cm) from the ceiling. If the air flow in your room draws in kitchen air, try to install that detector out of the path of those currents. Water vapor can also "trick" an ionization detector, so don't put it too close to bathrooms. The installation of a battery-powered smoke detector will be about the simplest household chore you will ever perform. No special tools are required. On the other hand, it's best to leave the installation of a hydro-powered detector to a professional.

For somebody in the alarm market, there is one more kind of detector to think about: a carbon monoxide gas sensor. A gas sensor could save your life by letting you know if your fireplace, your woodstove, propane-operated appliance or a malfunctioning furnace is sending out deadly carbon monoxide fumes. Like smoke detectors, these battery-powered units give off a low-battery warning but, unlike smoke detectors, they emit a beeping sound instead of a whining alarm. The possibility of carbon monoxide poisoning is a real concern for people living in modern energy-efficient homes, which are prone to chimney back-drafting—the air removed from the home via exhaust fans, clothes dryers, fireplaces and stoves can be replaced by air full of carbon monoxide.

Through all your deliberations regarding where to put which alarm, do not hesitate to contact your local fire department or home-security specialist for advice. Fire departments everywhere are glad to make suggestions.

Either an ionization or photoelectric smoke alarm can be connected to a siren or a horn on the exterior of your house. And the various alarms can be interconnected so that if the alarm in the basement sounds, it will set off the alarm on the third floor near the bedrooms. Lastly, if you have a centrally monitored security system, it ought to include fire alarm service, so that the central station will contact the fire department. The company should monitor your fire alarm system around the clock, even when your burglar alarm system is switched off. Your fire-protection system, when connected through your central monitoring station, helps ensure that the closest fire department is advised of an alarm from your home.

A More Systematic Approach

Detectors, of course, are working if they let you know about a fire. Your next concern is what to do about that fire. Because a household can be consumed by fire in a matter of minutes, it is imperative to plan in advance. You're not going to want to hold family meetings to discuss your next move while the fire alarm is blaring.

Every household should have a fire-safety contingency plan. This means holding regular drills. The local fire and police departments' emergency phone numbers should be posted clearly beside every telephone in the house, and train your family to use the numbers right away if they suspect fire. (Many Canadian urban centers now have an emergency "911" number.)

Draw a rough blueprint of your house, with all the rooms, windows, stairways and doors clearly illustrated. Make sure everyone knows of at least two exits from every room in the house. In a fire, one exit might be too dangerous to use.

When installing screens, storm windows or security bars, use only the kind that can be removed quickly from the inside. If any bedrooms have windows that are too high to safely jump out of, ensure that the room contains a rope or chain ladder that can be hung out the window.

If your family includes elderly or very young members, assign other members to serve as "buddies" in time of emergency. Train everyone in the use of extinguishers and other firefighting equipment. And—this is vitally important—arrange in advance to meet at a predetermined location outside the house should evacuation ever be necessary. Lives are lost when people re-enter a burning home looking for someone who is already outside and safe.

Anyone who becomes trapped in a room must keep the door closed, use blankets or towels to stop smoke from coming under the door, open a window and stay close to the fresh air near the floor. He or she should hang a sheet or pillowslip out the window to let firefighters know their whereabouts.

Parents of small children should take extra care to ensure that babysitters will know how to respond in a fire emergency. A babysitter's first responsibility, regardless of how small the fire appears to be, is always to remove the children from danger. The fire department can be called from outside the house, after the kids are evacuated. Forget about dressing them properly—wrapping them in blankets will be adequate. Babysitters should also be told that if they think they smell gas fumes, they should not turn on any electrical switches and should immediately call the fire department. They should know how to contact the parents, another responsible adult, police, firefighters and the ambulance. Post these numbers by the phone, for them and for yourself.

Apartment dwellers should know where the fire alarms and exits are located in their buildings. Never use elevators, and always feel a door before opening it. If it is hot, don't open it. Avoid smoke-filled corridors and close any doors between yourself and the fire. In either a house or an apartment, bedroom and hall doors should be kept firmly closed at night. If all that planning sounds like a pretty depressing topic for a family discussion, it should, because you're not going to have time if someone wakes up in the middle of the night yelling "FIRE!"

Fire Extinguishers

It is important to obtain and understand how to use various fire-extinguishing equipment. Proper use of a household fire extinguisher can mean the difference between a small, inexpensive problem and a huge, deadly conflagration. Fire extinguishers are classified according to their size and the types of fires they can effectively be used against. Class A fires include those fueled by wood, paper, textiles, rubber and several kinds of plastic. A Class B fire is fed by flammable liquids, gases, oils and greases. Class C fires involve live electrical equipment like wiring, motors and stoves. The other material burning in a Class C fire can be from the Class A or B categories. An extinguisher rated for Class C fires indicates that the extinguisher agent will not conduct electricity.

Modern extinguishers are made of aluminum or steel with a hand-operated valve lever at the top. A pin secures the handle in place and a breakable seal prevents accidental discharge. Most have either a pressure gauge or a pressure-testing button; clearly marked gauges provide a more accurate indication of the extinguisher's pressure level. Extinguishers can be safely

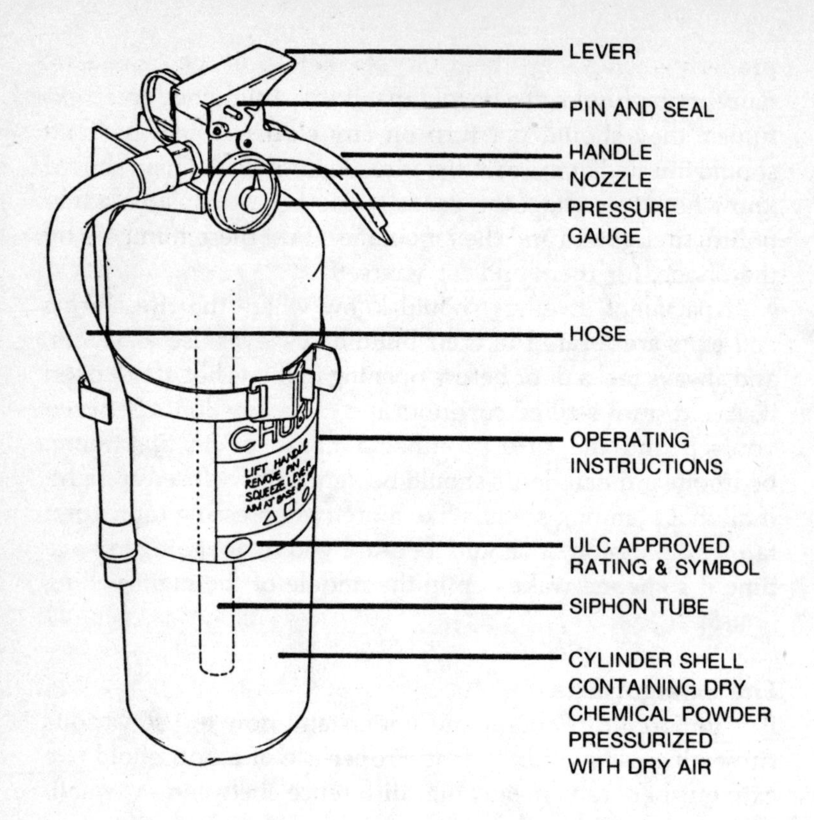

A TYPICAL HOME FIRE EXTINGUISHER

mounted on a bracket on a wall or in a vehicle. They're lightweight enough for a child to use, and it is essential that everybody in your home, including children, knows how to use them properly.

Fire extinguishers are loaded with a variety of extinguishing agents. Two very common dry chemical agents are monoammonium phosphate and sodium bicarbonate. The first is a multipurpose dry chemical and is effective on fires in all the categories, Classes A, B and C. The sodium bicarbonate chemical works effectively on Class B fires (flammable liquids). This type of extinguisher is less expensive than the monoammonium phosphate extinguisher, but is ineffective on wood, paper, cloth and rubber fires.

Extinguishers can also contain carbon dioxide, which is suitable for Class B fires, or pressurized water, to be used only on Class A fires. Neither of these is suitable for a home environment.

For extra security, there are spray bottles with liquid flame retardant that you can spray over areas near potential fire hazards, such as fireplaces, woodstoves, or on your Christmas tree.

For household use, certain sizes and styles of extinguishers are more effective than others. First of all, any extinguisher you own should carry an Underwriters' Laboratories of Canada rating. You might be able to find a very cheap, non-ULC-rated extinguisher, but the risk that it may not work in an emergency is not worth the lower price. The ULC rating will tell you specifically on which types of fires the extinguisher works most effectively. The average household is best served by a multipurpose monoammonium extinguisher, which is marked by the letters A, B and C. This versatile extinguisher is best kept in the kitchen, where about 65 percent of household fires start. Other good locations are the workshop, garage and furnace room. Ideally, your home will have extinguishers near escape routes, doorways, stairwells and hall closets; you should also keep them in cars, trucks, RVs and power boats. Don't hide them in drawers, cupboards, or on high shelves, where they could remain out of sight and forgotten.

Most high-quality fire extinguishers are rechargeable, although companies are now producing effective disposable extinguishers. If you ever do use a rechargeable extinguisher, have it recharged by the dealer as soon as possible. Even if some powder is left in the unit, it won't hold its pressure charge and would be useless in fighting another fire.

Fire extinguishers require minimum maintenance but every month, give them a visual inspection — make sure the pin is still in place and the pressure is strong.

When it comes time to use an extinguisher, treat the situation as an emergency. Yell "Fire!" so people will clear out of the area and call the Fire Department. Then, if you think it is safe to do so, and the fire is not too big, grab the extinguisher, remove the safety pin, grip the handle firmly and aim at the base of the flames. Then, from between 10 and 12 feet (2-4 m) away from the flames, squirt the contents onto the fire with a sweeping side-to-side motion. If you can't approach within 12 feet (4 m) of the fire, it's too big and you should forget about fighting it. Get out of the house. However, if you can extinguish the flames, once you are certain they are out, ventilate the area quickly to avoid dangerous smoke or fumes. Then call the fire department and let them know exactly what happened. They may want to inspect the property for smoldering materials or other hazards that could reignite the fire.

Another firefighting tool for domestic use is the fire blanket. Usually packaged in a roll-up container that can be hung on a wall, a fire blanket is woven of fire-resistant, non-toxic glass fiber. When used properly, a fire blanket smothers small domestic fires by cutting off the oxygen supply to them. They are reusable, easy-to-use effective tools for the workshop, kitchen and garage, but they are not designed to replace fire extinguishers — only to complement them.

When a fire does get too big for an attack with either an extinguisher or a blanket, get out. Although you can have fun at a party by playing fire and imagining which household items you would be most likely to save in the event of a fire, there is no time to do so when the real thing strikes. And this should be stressed during the family fire drills. No toys, no money, no jewelry, not even the family pet — you and your loved ones are the only concerns.

As stated in Chapter Three, if you are thinking about buying a fire-resistant safe, be sure you don't confuse it with a burglar-resistant safe, although some companies manufacture dual-purpose safes, with a fire-resistant safe built into a burglar-resistant one. However, safes are clearly labeled "fire-resistant" and not "fireproof" for a reason. Steer clear of any salesperson who offers you a completely fireproof safe. High-quality fire-resistant safes are sound investments, but the completely fireproof safe has yet to be built. Good fire-resistant safes are tested by Underwriters' Laboratories to withstand at least one half hour of 1560°F (850°C) and are assigned a rating accordingly. A worthwhile safe is water-resistant and durable, and tested for falling in case the floor collapses. A good fire-resistant safe can keep important papers, small items and sensitive items such as computer disks safe from heat.

Any number of these pieces of equipment can help to protect you, your family and your home in the event of a fire. But remember what your greatest defense is: the best fire-resistant safe, the most elaborate fire extinguisher and the loudest alarm guarantee no security unless they are mixed with healthy doses of common sense and household care.

CHAPTER EIGHT

WORKING WITH A SAFETY NET: INSURANCE

"The day after the robbery, I phoned the insurance company and an adjustor came over. I had insurance on everything, but not on the jewelry. Most policies cover up to $2,000 on what they call 'unspecified jewelry.' And even though I don't wear jewelry — I find it an encumbrance — we had acquired quite a bit. There was some quite beautiful stuff, with tons of emotional value. And it was all there just waiting to be taken. And you think now of course it should have been in a safety deposit box. But it wasn't, and just one bracelet, which was woven with three different kinds of gold, was worth about $2,000.

"Mostly, the adjustor simply took my word that we had had that amount of jewelry. After all, the jewelry cases were all there, turned over. But he did want evidence that all the stereo equipment was stolen. He wanted to see the instruction booklets or the warranty covers. In one case I had a thing in my wallet from Canada Customs with the serial number of one of my cameras. I guess I'm the kind of person who keeps that kind of information around, fortunately.

"The aggravating thing is that you sometimes get objects taken or things stolen that are no longer replaceable because they're not made anymore. The very small 35-mm camera I had taken is no longer made. And although my insurance is replacement value, you do have to replace it. For instance, I had a hi-fi VCR, top of the line. They will replace it with another top-of-the-line hi-fi VCR of comparable value. At list price. If mine is worth $950, they give me $950 toward a new one. If you choose not to replace it, they will figure a value based on the depreciated value.

"In total, I lost somewhere around $12,000 worth of goods, including the jewelry, hi-fi equipment, camera and watches. Of that, about $8,000 was in unspecified jewelry, so I only got back $6,000."

In the insurance industry, some people refer to a sales pitch that uses scare tactics to encourage people to buy insurance as "backing the hearse to the door." It's no longer a popular tactic, fortunately. An authoritative book by James Fleming on the corporate side of the Canadian insurance industry is called *Merchants of Fear* (Viking, 1986), a title that reflects a popular but largely unfounded image of the insurance sales business.

Lest you think property insurance companies must "back the hearse to the door" in an effort to scare you into buying property insurance, stop for a moment and look at the average cost of a standard household and its theft insurance premium. Compared to the total value of all your insurable goods, the premium is quite low. There should be very little need for persuasion. Depending of course on your personal circumstances, an average householder premium amounts to about $250 a year. That will insure your home against:

- Fire;
- Theft;
- Liability in case someone falls down on your sidewalk and takes you to court;
- Third-party liability in case your son or daughter breaks a neighbor's window with a football;

- Protection for when you are away from home; and
- A wide assortment of other problems.

A quick assessment of your household property—stereo, camera equipment, silverware, jewelry, coin collection, art, etc. — will show you that the insurance premium pays for itself pretty quickly if you, like millions of other Canadians, fall victim to either a blaze or a burglar. Ask yourself how much that premium really costs when compared to the total worth of your property.

Today, in what has been called the most heavily insured country in the world—Canada—more than $10 billion worth of policies are written yearly. More than a third of that is property insurance, with a majority of the rest going toward automobile insurance. In 1988, the average household was spending about $1,750 annually on security, a figure comprising life, auto and property insurance.

Canadian insurance experts say that a majority of policyholders underestimate their true material worth. The average home contains goods worth about 60 percent of the building value. In other words, if your home is valued at $200,000 (not including the land), then you probably have nearly $120,000 worth of goods within. If you are doubtful of that figure, remember that it includes everything—furniture, appliances, clothes, toys, tools, jewelry, collections, art, hobby materials, home-entertainment gear, books, records, musical instruments, cameras, sporting goods, coin collections; it adds up quickly. So you may find your policy covers only so much and that it is time to reassess your valuables. (See the Checklist on Insurance in the Afterword.)

Unlike homeowners, apartment dwellers are notorious for neglecting property insurance until after some disaster strikes them. Unless they carry their own household insurance, apartment dwellers have no protection against theft. Regardless of how much insurance the landlord carries, if you are a renter, you must arrange your own policies. Proper insurance not only protects you if you get broken into and robbed but it also pays the bills if something that happens in your apartment — a bathtub flood, for instance — causes damage in somebody else's apartment. Liability insurance also protects you if someone is injured in your premises and takes you to court. Furthermore, a carefully developed tenant's insurance package will insure you against mishaps that occur away from the home and losses that other people suffer around your property.

As with any consumer product, the range of insurance policies is quite varied. Inevitably, the questions arise: What kind of insurance do you need and how much of it should you purchase? So before knocking on your local agent's door, consider the kinds of protection available.

Personal Property

First, a question: If you were to return home some evening to an empty, looted house, and then the following day be asked to put a monetary value on everything that had been taken, would you be able to come up with an accurate figure? How about this one: Do you know who foots the bill if your furniture gets damaged while you are moving from one house to another? What about pets — can you insure the family dog? Insuring your personal property appears easy, until you take a closer look. Often the task of figuring out what is covered and what isn't covered is confusing. What if you brought some work-related equipment home from the office and something happened to it while it was in your house? Does your personal-property insurance policy cover things like that?

In general, most people have only a layperson's knowledge of how their personal property is insured under ordinary everyday circumstances. More knowledge, and often more insurance, would be better.

Personal property includes everything you own, except for vehicles — your car, motorcycle, airplane, jeep or motor boat, which must be insured separately. Into the personal-property category, we include:

- clothes
- appliances
- computer equipment

- jewelry
- collectibles
- cameras
- photo collections
- stereo gear
- furniture
- sporting goods
- books, records and
- anything else you own that is valuable.

The dollar value of these items, not their sentimental value or size, dictates how much insurance you should carry. It's one thing to count up the items you own, and quite another to assess their worth. To make the job of figuring out the worth of your goods easier, you can obtain an inventory list from your local insurance agent.

A basic personal-property insurance plan, be it part of a simple fire policy, homeowner's policy or a tenant's or condominium owner's plan, will cover property losses only up to a certain maximum. Beyond that maximum, you should talk with your agent about a "floater," which buoys up your coverage so you would receive proper compensation for specific lost items. A floater, or "rider," should be purchased if you are insuring commodities such as:

- coin collections
- stamp collections
- bullion
- bank notes
- securities
- manuscripts
- computer data
- jewelry
- furs
- antiques
- artwork
- musical instruments

A standard policy without a floater will cover the loss of these items, but only to a mutually agreed-upon maximum amount. If you have more than $2,000 worth of jewelry sitting around your house (and you would be surprised how easy it is for a small personal collection to exceed that amount), you should think about a floater. It will go a long way toward keeping your most prized property adequately insured.

You must also consider whether or not you want "replacement cost" insurance. With replacement cost, your goods will be valued at the price it would cost to replace them at the time of their theft or destruction by fire, with no depreciation from the moment you acquired them. Otherwise, your insurance policy will provide coverage for only the original cost of an item minus its depreciation since you acquired it. Your stereo system might have cost $3,000 new but carries a resale value of less than $1,000. Without replacement cost insurance, you will recover only the lower of the two prices.

For personal peace of mind, it is important to know what you own, what it is worth and whether or not it is covered in your insurance policy. If you purchase a $1,000 pair of stereo speakers, make sure you find out if you have to extend your policy to cover them.

Every now and then, take inventory. At the same time, you might want to take photographs of your property. Each picture would be worth a thousand words, and perhaps thousands of dollars when it comes to both insurance payouts and police investigations. Better still is a video record, because you can film your belongings from all angles, describing them aloud at the same time. (You might want to make imaginative use of mirrors in order to film the video camera itself, if it's yours.) Videotapes can also be updated very economically and quickly. One woman with an extensive book collection records the title and price of every new book she acquires, so the insurance company will be able to estimate the collection's worth. Another man, who lost a large collection of photographs in a fire, was able to replace every single snapshot because he had had the foresight to keep the negatives. Of even greater significance, he kept his negatives in a safety deposit box, away from the house.

If you are going to make such a record, keep it safe from loss. A thief who grabs your VCR might also take the tapes lying beside it, and if your homemade inventory videotape is among them, you won't have much to show the insurance adjustor when he or she visits. If you can't store lists, receipts, warranties, tapes, photos and other documentation in a fire-resistant or burglar-resistant safe, then they should be stored away from your home. They can be replaced more easily than the goods they represent.

For more highly valued homes — those that cost more than $500,000 — an insurance company might hire a professional residential inspection company to conduct a thorough examination and evaluation of the property. The evaluators will assess the replacement cost of the house and suggest what steps can be taken to increase the degree of burglary and fire protection before the home is insured. As the owner of a high-valued home, you might want to engage these evaluators to measure the worth of your property. And you can count on the final figures being a surprise. If you are interested, ask your insurance agent for more details on the evaluators, known as the Insurers Advisory Organization.

After you have ascertained and documented your belongings, find out whether or not they can be insured. Sometimes an item is not insured once it is removed from your home. Likewise, your furniture might not be covered during a move from one dwelling to the next. Pets can be insured, but not if they are stolen or hit by a car. Your insurance might take effect if a visitor to your home loses a possession there, but be sure you know how such situations are spelled out in your policy.

Fire Insurance

When an insurance company representative discusses fire insurance premiums with you, the coverage is usually presented as part of a larger homeowner's policy, a policy that includes theft and liability. Nevertheless, some companies will provide simple, straightforward fire insurance, without the accompanying coverage. Even simple fire insurance protects you from loss or damage caused by a variety of perils: explosions, falling objects such as airborne items or tree limbs, vehicles that crash into your building or property, lightning, and damage done by vandals, hurricanes or tornadoes, to name a few.

Like every other kind of insurance rate, the amount you pay for fire-insurance premiums is determined by odds. Insurers look at the value of your property, your location in relation to fire hydrants and fire stations, how well or poorly equipped the fire department is, your neighborhood and, finally, the frequency of fires in your area. Some insurance companies will want to send around a representative to calculate your premium and determine the replacement cost of your building and contents. Replacement cost is not the price you paid for your home because that included the lot, which can never burn down.

The assessor will note such variables as building material, number of stories, state of the attic and basement, location and condition of fireplaces, porches, garages, amenities (such as air conditioning), and design (is your home detached, semi-detached or a row house?). You can make your own estimate using the Insurance Bureau of Canada guidelines. And unless you are charged with arson, your insurance policy will recover the full cost of the building and about 60 percent of its contents.

Liability Insurance

We have all heard the apocryphal tale of the thief who is in the middle of breaking into a home when he is startled by the homeowner. Making a break for it, he dashes across the porch, down the stairs and right onto a patch of ice, where he slips and breaks a limb. He then sues the homeowner for negligence and wins.

Fortunately, it never happened. Not in Canada, at any rate.

But because we live in a society where people are more likely than ever before to take you to court, personal liability deserves priority among your insurance considerations. Between 1985 and 1986, the number of liability insurance policies written in Canada increased an incredible 63 percent. If you think about the prospects for a moment, you'll quickly realize how easily you can be held accountable for innumerable possible mishaps, including another person's property damage, personal injury or even death. For example, imagine a delivery person slipping on your steps and injuring herself seriously enough to sue. Because a policy protects the policyholder's immediate family as well, if your son or daughter accidentally hurts a playmate and the child's parents hold you financially responsible, adequate liability insurance will ensure that you don't go broke paying the settlement. If you carry no such insurance, you might wind up paying the cost of that person's personal losses for the rest of your life.

Liability extends to property damage as well. If a fire originating in your back yard burns down your neighbor's shed, you could be found legally responsible. In the same vein, something called fire legal liability insurance protects homeowners and tenants against having to pay for other people's property losses that otherwise would be excluded under personal liability insurance because the premises are rented. Even if a fire in your apartment is self-contained, if it affects other people's property, or if you have a fire in a hotel room or other rented accommodation, you might be asked to pay the repair bills. Fire legal liability covers that cost.

If you are in a position in which your actions or those of your family might cause other people to experience loss or injury because of some small mistake, liability insurance is virtually obligatory. Do you have domestic workers in your home? Do you ever drive a snowmobile or other non-licensed vehicle? If you so much as let a local kid cut your grass with your lawnmower, you are looking at a liability risk. Even if you cannot be held legally liable for actions that cause other people harm, you might be taken to court where you and your lawyers will have to prove your innocence. And somebody has to pay the legal costs. Appropriate liability insurance will do so.

The Insurance Menu

Like the Ford motorcar, home insurance packages at one time came in one variety: fire. Now you and your home can be insured against the following: fire, lightning, windstorms, hail, explosions, riot damage, aircraft damage, damage caused by a vehicle operated by the policyholder, smoke, vandalism, glass breakage, theft, falling objects, snow or ice damage, building collapse, plumbing and water-pipe problems, and damage from water or accidentally generated current to appliances, fixtures and wires.

Like the auto industry, the insurance industry is comprised of many competing companies, and their policies and offerings differ from one another in quite a few ways. Wording will vary, and some policies will offer extras, like credit-card insurance, as a bonus. Forearmed with a precise assessment of your own needs, however, you can discuss your precise insurance requirements with some authority when it comes to actually buying a household policy for yourself.

Essentially, you will be faced with a choice of four types of policies: fire insurance, which is seldom sold on its own these days; a standard homeowner's package; a comparable tenant's package; and the condominium-unit owner's package.

As you might guess, fire insurance offers coverage in the event of fire and a handful of other dangers, but offers little help to people who lose their belongings in a robbery. More practical for a homeowner is a variation on the standard homeowner's package, which includes fire with theft and liability, all covered under one premium. If you want coverage for other belongings, like a cottage, a boat or other valuable property, you will have to add a floater to the standard homeowner's package. Or, if you prefer a lower premium (and less protection), you might want to ask about a limited homeowner's package, which would not include theft coverage. Keep in mind that homeowners pay more for home and contents insurance if they are frequent travelers, live in high-vandalism-and-arson areas such as slums or luxury neighborhoods, or in remote areas far from fire halls.

A tenant's package is similar to the homeowner's but excludes any coverage of the building itself, which should be insured by the building's owner. Too often, tenants assume that a landlord carries insurance covering a tenant's loss, but that is simply not the case. So, insurance companies offer packages that not only protect your personal property but also offer legal liability insurance, in case something you do in your apartment affects your neighbors. You might find that for every thousand dollars' worth of insurance, a tenant's policy is more expensive than a standard homeowner's package, but that is because insurance companies have calculated that tenants are higher risks, particularly those in urban areas.

Condominiums have insurance rules all of their own. In theory, condominium owners are in charge of the building manager, and it is their responsibility to make sure the manager buys insurance for the building and common areas. Then the owners have the option of buying a condo-unit owner's package, which is similar to a tenant's package except that it includes coverage for aspects of condo life that would not apply to regular tenants. For one thing, a condominium-unit owner's package will provide up to $1,000 coverage for improvements the owner has made to the unit and up to $5,000 insurance in case the condominium corporation loses money and everybody is forced to pay. And if the corporation is made liable for any costs for which it is not insured, the condominium-unit owner's insurance policy will pay that owner's share of the costs, up to $5,000 with a standard policy.

All four packages cover the immediate family and the immediate family's property. That means wife, husband and children. Although the law recognizes common-law relationships, you would be well advised to explicitly state your mate's name on your policy if you are living together without having tied the knot.

Furthermore, if you have taken any extra anti-theft or

anti-fire precautions, let your agent know about them. For example, if your dwelling is equipped with a burglary and fire alarm system that notifies a central monitoring station, you will be eligible for a reduction in your insurance rates. Also, if you have taken steps to guard your property against intruders, or if you are a participant in a building program like the Shield of Confidence, make sure your agent knows. Some give discounts if you have fire extinguishers, deadbolt locks and similar low-cost security measures. Discounts might also be offered to non-smokers and non-drinkers.

What To Do When the Time Comes . . .

Unfortunately, the odds are good that some day you are going to have to make a claim on your insurance policy. If you do come home to a looted or burned house or apartment, or if you have reason to believe that someone is about to hold you financially responsible for some unforeseen problem, contact your insurance agency as soon as you can.

With regard to break-ins, follow these guidelines: Upon your arrival home, if anything about your house looks suspicious, don't go in. Go to a neighbor's and call the police. An intruder might be armed and dangerous, especially if caught in the act. If everything looks fine from the outside but you encounter a burglar inside the house, try to avoid a confrontation. And try not to panic. You might get a chance to push a panic button (if you have one) that would alert a central station to your problem. Make mental notes about the thief's appearance — clothing, coloring, height, weight, unique characteristics — and let him leave. Then phone the police and give them your full name and address. Don't touch anything. When the police arrive, tell them every detail you can about whom you saw and what you lost.

Then call your insurance agent. The agent will dispatch an adjustor to verify your losses and begin the task of assessing your coverage. After the adjustor has determined the loss, he or she has 60 days in which to settle. If you have a record of all your belongings and their respective values, the adjustor's job will be much easier and you will be able to replace your goods expeditiously. If you don't have any record, draw one up from memory as soon as you can. Don't fret if you miss a couple of items when you first talk with the adjustor; they are generally reasonable people who will take your word for it when you phone them a few days later and tell them you forgot to include an item or two. Insurance experts understand that they almost always deal with claims in time of trauma, so it is no sin to be a bit forgetful.

Depending on your policy, the insurance company will offer you the replacement value or the actual value of the items that have disappeared. Sadly, it is impossible to assess the sentimental value of lost goods, so you can only expect to receive market value of your inherited pocket watch. There is no reason to be suspicious if your agent recommends a store or manufacturer who can replace your goods at a discount. Insurance companies sometimes obtain legitimate bulk-volume deals that the general public would not otherwise be able to get.

Throughout this unpleasant but necessary process, don't hesitate to discuss with your agent any doubts or desires you have about the ordeal. Remember, most insurance claims are handled in a smooth and efficient manner. It's only the sensational and unusual ones that you read about in the newspaper. Finally, remember that the uninsured homeowner or tenant suffers much more grievously when he or she is hit by a robbery, fire or lawsuit.

AFTERWORD

SECURITY: IT'S HABIT FORMING

"You become very conscientious. Day to day, I'm very apprehensive. Mind you, now I have a timer operating a radio very loud during the day. Now I check the timers when I go away. I have four timers cycling lights on and off in the den by the TV room and the light shines through the shades so you can't see into the room from outside. My bedroom light cycles on and off.

"There was no light at all on the north side of the house where the window was. Plus there was this big bush. I couldn't fit bars on the window that's vulnerable, so I put Lexan on it, this high-tech plastic stuff.

"On the north side of the house I've got two big spotlights with a motion sensor, so if you come anywhere near that side of the house, the lights come on for about 20 minutes and then they go off.

"I now phone people to check the house when I'm away, and when I was away in January I had a friend over to shovel the walk because that's a real clue. The house looks lived in when you go by it now. If it snows during the night, I shovel the walk before I go to work in the morning.

"In any event, the entire affair has made me very cynical. There is that victimized feeling that you go around with for a long time. Life is difficult enough, and this is something you don't need."

It is a rare Canadian who hops into his car and roars off without buckling the seatbelt. Even more difficult to find is a young or even professional hockey player leaping over the boards for some on-ice combat without first donning protective headgear. And few contractors would consider erecting a new house without installing the requisite number of smoke detectors. In our continuing quest for peace of mind, security and prosperity, Canadians have shown admirable readiness to embrace ideas that make life safer.

In addition to applying good old-fashioned common sense to how we drive, play our games and build our homes, it is beneficial to remember that not too many years ago, little things that we today consider necessities of life were deemed as unnecessary as a snowplow in July. For example, young drivers often failed to carry public liability insurance. They sure do today. And for that matter, what physician, however reluctantly, neglects to pay malpractice premiums? Most of us lock our car doors, and some join Neighbourhood Watch programs. And although we almost all lead law-abiding lives as citizens, we take precautions against having taken from us those things we work so diligently to acquire. And we take those precautions for granted.

It is unfortunate that we have to concern ourselves with personal safety, other people's liability and generally falling innocent prey to forces beyond our control. But the world is changing and we must adapt, just as we have with seatbelts and hockey helmets. Adequate insurance against what sometimes seems like an unforgiving outside world allows us to sleep better at night and make the most of the days.

But in learning about home security, about hanging on to your worldly goods with minimal anxiety and ensuring the safety of your loved ones, it is imperative to approach the subject optimistically. Just because you want to secure your property against criminals and fire doesn't mean you have to nurture anxiety or paranoia. The aim should be to increase your ease-of-living, not to detract from it.

Clearly, the individual who ought to fret about his home-security system is the person who lives in an unsecured, easily entered dwelling. Just ask any of the nearly half million Canadians whose lives were sadly darkened by a break-in recently. And being broken into and robbed is a scarring, unforgettable experience that, yes, feels like rape. Security is a fundamental human right, and when it is violated by some stranger, the experience is truly nightmarish.

The idea would be to live with a home-security system as comfortably as we have learned to live with seatbelts. When you strap yourself into your car, you're not thinking about car accidents. You don't consciously recall those horrifying seatbelt commercials that show you what can happen if you find yourself in an emergency situation without a belt. That would be all too unnerving, too worrisome. All you really do is strap yourself in and enjoy the drive. Similarly, if you live in a safe environment, you need not worry about thieves or flames. All you have to worry about is how to enjoy those things you have worked so hard to attain, including peace of mind.

The very fact that you are reading these words shows that you are interested in establishing a more secure, comfortable life. The first step in increasing your home and personal security is to gather information and determine what's best for you. Because all Canadians, from St. John's to Victoria, are interested in fighting crime and increasing the quality of their lives, government agencies and many public service groups are eager to work with individuals on projects that lower crime and disaster statistics. So don't hesitate to talk with other people about your security concerns and become involved in community projects. The answer lies in everybody working together.

The following checklists provide a quick reference for you and your family. Go through them and see if there are any steps you can take today to make your life more enjoyable.

HOME SECURITY QUIZ

The following is a quick test to measure your home's susceptibility to break-ins. To see if you are a likely candidate for a break-and-enter, add together the numbers that apply to the description of your home. How does your house rate?

House Design

Type of House:
(6) Detached
(3) Semi-detached
(3) Townhouse

Configuration:
(3) One-story
(2) Two-story
(2) Three-story

Approximate Area:
(8) More than 3000 sq. ft. (270 m^2)
(4) 2000-3000 sq. ft. (180-270 m^2)
(2) under 2000 sq. ft. (180 m^2)

Entrance Doors:
(4) Sliding glass (each)
(3) Hinged single (each)
(3) Hinged double (each)
(2) Garage into house

Windows:
Ground Floor (2) Opening (each)
(1) Fixed (each)
Basement (3) Opening (each)
(2) Fixed (each)

Upstairs Window(s):
Accessible from garage, roof, balcony or other structure?
(1) Yes

Exterior Illumination:
(4) None
(2) Door areas only

Immediate Environment

Distance to Nearest Neighbor:
(4) More than 100 ft. (30 m)
(2) 25-100 ft. (8-30 m)
(1) Less than 25 ft. (8 m)

Abutting Property:
(8) Ravine or woodlot (6) Golf course, field or other open space (1) Neighbor's yard
Trees and Large Shrubs on Property? (2) Many (1) Few
Windows Concealed by Shrubs? (4) Yes
Entrance Doors Visible from Street? (2) No
Entrance Doors Visible by Neighbors? (2) No
Burglaries in the Immediate Area in the Last 12 Months?
(8) Three or more (4) One or two
Lifestyles

House Unoccupied:
(8) All day (4) At times during the day and often in the evening (2) During vacations

Nearest Neighbors:
(2) Away frequently (1) Away sometimes
Spouse Makes Overnight Business Trips:
(2) Often (1) Sometimes

Score Results:
0-30 Low Risk. Statistics show that homes that score within this range are less likely to be burglarized. That doesn't mean, however, that you should forget normal day-to-day security habits.
30-60 Medium Risk. Your home has many of the things burglars look for when casing a potential job. Improved security is recommended.
Over 60 High Risk. It's time you took a serious look at your property. Advanced-level security precautions are imperative.

Security Measures

This list can be used to check your present risk situation. Deduct the total of the numbers indicated from your score in the Home Security Quiz. It is also a handy checklist of the more important crime prevention measures you can apply to your home.

Lighting:
(2) Light timers
(4) Perimeter lights
(4) Motion-sensor lights

Dog:
(8) Guard dog
(3) Family pet

Doors and Frames:
(4) Steel
(2) Controlled key locks
(3) Patio slider—extra locks
(2) Removable thumb turns
(3) Deadbolt locks

Warning Signs:
(4) Neighbourhood Watch
(5) Operation Identification
(6) Alarm system
(6) Guard dog

Windows:
(5) Security bars: basement
(4) Locks

Security system:
(10) Centrally monitored alarm
(4) Local alarm

Checklist of Household Valuables

Do you have any of the following in your home? These items are a thief's favorite targets.

— Money
— Guns
— Jewelry
— Coin/Stamp collections
— Original artwork
— Antiques or irreplaceable objects
— Wine cellar
— Watches
— Furs
— Stocks and bonds
— Silverware or silver service
— Collectibles (figurines, etc.)
— Cameras
— Stereo
— VCR
— Television
— Personal computer
— Other high-value items

Checklist on Fire Prevention

If you answer "no" to any of the following, your home could be in danger of severe fire losses.

(1) Is your home equipped with working smoke detectors on every floor?

(2) Have you multipurpose fire extinguishers in the kitchen, garage and workshop?

(3) Have your extinguishers been inspected according to the manufacturer's instructions?

(4) Is there a fire blanket in the kitchen?

(5) Has your fireplace/woodstove chimney been professionally cleaned within the past year?

(6) Is your fireplace well screened?

(7) Does your woodstove meet Canadian Standards Association installation requirements?

(8) Do you have your furnace cleaned and inspected annually?

(9) Are all your electrical circuits used only to their legal capacity?

(10) Do you make sure flammable liquids are used only outdoors?

(11) Is the area around your furnace kept free of combustible materials?

(12) Do you have a fire escape plan and do you regularly review fire-safety practices with your family?

(13) Does everyone in your home know the fire emergency phone number?

Checklist on Insurance

In the event of loss, how well protected are you?

Do you carry:
(1) Homeowner's insurance?
(2) Tenant's insurance?
(3) Condominium insurance?
(4) Adequate liability for household employees?
(5) No insurance?

Does your policy:
(1) Pay replacement value?
(2) Pay depreciated value?
(3) Include a reasonable floater for valuables?
(4) Cover your away-from-home losses?
(5) Reflect the true recently assessed value of your home?

Have you kept an inventory of your belongings in a safe place?